WAYS TO READ

IN PRINT

T0261062

Make: DIY Heroes

Off the Hook **Rotary Cell Phone**

Makers Step Up to Combat Covid-19
- **Community PPE Guide**
- **Self-Sufficiency:** Homeschooling Homesteading Sourcing Materials
- **Free and Easy Webcasting**
- **Build a 20-Second Musical Hand Soap Dispenser**

5 Family Maker Camp Projects to Save Your Sanity

39 PROJECTS: Backyard Wind Turbine Upcycling Old PCs Mini Jacob's Ladder Sequin Rewrite Clock

makezine.com | make.co | makerfaire.com   VOLUME 73

Calling All Climate Makers: **Electrify Everything!**

"Inner Glow" LED Heart
Create brilliant displays and décor with a new edge-lighting trick

ON THE GO

ONLINE

Create incredible projects, learn useful skills, and meet the inspiring maker community...
Subscribe to the premier DIY magazine now!

OVER 120 PAGES OF:
- Detailed step-by-step projects
- Tips and skill-building tutorials
- Cutting edge tool reviews and recommendations
- Everything from high-tech builds to artisanal crafts

Subscribe to *Make:* at makezine.com/go/subscribe
Become a Make: Community member at make.co

Subscribers and members can access the digital magazine (and our library with thousands of projects) on the web and in the iOS and Android apps

CONTENTS

ON THE COVER:
The Adafruit Clue microcontroller lights up with a graphic from the PyCon 2020 conference.
Credits: Adafruit, Robb Design/PyCon

24

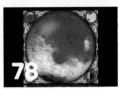

Make:

> "Please don't start making videos because you want to become rich and famous. Make videos because you want to share your passion, and because you want to get better at making videos, and you want to get better at making things." –Mark Rober

PRESIDENT
Dale Dougherty
dale@make.co

VP, PARTNERSHIPS
Todd Sotkiewicz
todd@make.co

EDITORIAL

EXECUTIVE EDITOR
Mike Senese
mike@make.co

SENIOR EDITORS
Keith Hammond
keith@make.co

Caleb Kraft
caleb@make.co

PRODUCTION MANAGER
Craig Couden

CONTRIBUTING EDITOR
William Gurstelle

CONTRIBUTING WRITERS
Sam Brown, Colton Coty, Sarah Fogle, Joey Freid, Kathy Giori, Saul Griffith, Mel Ho, Kris Kepler, Bob Knetzger, Helen Leigh, Eleanor Lutz, Mimi the Nerd, Forrest M. Mims III, Pi-Hole Dev Team, Samantha Reardon, Jane Stewart, Tim Sway, John Thurmond, Wesley Treat, Dr. Mel Vallero, Dr. Randell Vallero, Adam Vera, Daniel West

DESIGN & PHOTOGRAPHY

CREATIVE DIRECTOR
Juliann Brown

CONTRIBUTING ARTISTS
Caleb Kraft

MAKE.CO

ENGINEERING MANAGER
Alicia Williams

WEB APPLICATION DEVELOPER
Rio Roth-Barreiro

GLOBAL MAKER FAIRE

MANAGING DIRECTOR, GLOBAL MAKER FAIRE
Katie D. Kunde

MAKER RELATIONS
Sianna Alcorn

GLOBAL LICENSING
Jennifer Blakeslee

BOOKS

BOOKS EDITOR
Patrick DiJusto

MARKETING

DIRECTOR OF MARKETING
Gillian Mutti

OPERATIONS

ADMINISTRATIVE MANAGER
Cathy Shanahan

ACCOUNTING MANAGER
Kelly Marshall

OPERATIONS MANAGER & MAKER SHED
Rob Bullington

PUBLISHED BY

MAKE COMMUNITY, LLC
Dale Dougherty

Copyright © 2020
Make Community, LLC. All rights reserved. Reproduction without permission is prohibited. Printed in the USA by Schumann Printers, Inc.

Comments may be sent to:
editor@makezine.com

Visit us online:
make.co

Follow us:
🐦 @make @makerfaire @makershed
f makemagazine
makemagazine
makemagazine
twitch.tv/make
makemagazine

Manage your account online, including change of address:
makezine.com/account
866-289-8847 toll-free
in U.S. and Canada
818-487-2037,
5 a.m.–5 p.m., PST
cs@readerservices.makezine.com

Make: Community

Support for the publication of *Make:* magazine is made possible in part by the members of Make: Community. Join us at make.co.

CONTRIBUTORS

What's your favorite maker-focused YouTube channel?

Helen Leigh
Berlin, Germany
(Python on Hardware)
Rich Rebuilds is a mechanic who rebuilds damaged Teslas and advocates for the right to repair, something all makers should get behind!

Daniel West
Wollongong, NSW, Australia
(Universal Lego Sorter)
Wintergatan — a musician and engineer documents his quest to build a groundbreaking machine that plays instruments with bouncing marbles.

Samantha Reardon
Oakland, California
(DIY Mobile Handwashing Station)
My favorite is *Styropyro* because of his incredible ability to transform household items into magically powerful items that leave the viewer feeling excited and stupefied.

Issue No. 74, Fall 2020. *Make:* (ISSN 1556-2336) is published quarterly by Make Community, LLC, in the months of February, May, Aug, and Nov. Make Community is located at 150 Todd Road, Suite 200, Santa Rosa, CA 95407. SUBSCRIPTIONS: Send all subscription requests to *Make:*, P.O. Box 17046, North Hollywood, CA 91615-9588 or subscribe online at makezine.com/offer or via phone at (866) 289-8847 (U.S. and Canada); all other countries call (818) 487-2037. Subscriptions are available for $34.99 for 1 year (4 issues) in the United States; in Canada: $43.99 USD; all other countries: $49.99 USD. Periodicals Postage Paid at San Francisco, CA, and at additional mailing offices. POSTMASTER: Send address changes to *Make:*, P.O. Box 17046, North Hollywood, CA 91615-9588. Canada Post Publications Mail Agreement Number 41129568. CANADA POSTMASTER: Send address changes to: Make Community, PO Box 456, Niagara Falls, ON L2E 6V2

PRINTED WITH **SOY INK**

Sharing Skills

by Dale Dougherty, *Make:* President

Bruce Sterling wrote about flintknapping — the making of stone tools — in the first issue of *Make:* in an article wonderfully titled "Make the Tools That Made You." He wrote that stone tools date back over two millions years. "Modern humans are just 200,000 years old" so our hominid "ancestors spent 1.8 million years making tools out of rocks."

Search for flintknapping on YouTube and you'll find many instructional videos, usually with a person showing their knicked-up hands directing a blow of one rock on another. Knapping is not something you can learn just by watching. You need to develop the feel for it, which requires hundreds of hours of practice.

Dietrich Stout, an experimental anthropologist at Emory University in Atlanta, spent 300 hours learning to make a stone hand axe. Now he's having his students learn knapping as well. Stout's interest is "what happens inside the brain when a person knaps away at a piece of stone." Stout and his colleagues did MRI scans of the students' brains to look for changes over time that might provide clues to human evolutionary development. "What matters is the kind of tools we make and how we learn to make them," he writes in a *Scientific American* article, "Stone Age Brains" (April 2016). Stout thinks "humans stand out in their ability to learn from one another."

The development of more complex stone tools coincides with the development in humans of larger brains with new capabilities. "The demands of toolmaking — combined with complex social interactions for teaching these skills — may have become driving forces for human cognitive evolution," writes Stout. "Teaching and learning increasingly complex toolmaking may even have posed a formidable enough challenge to our human ancestors that it spurred evolution of human language." There was an evolutionary advantage to being able to make and use complex tools as well as using language to share the learning process with others. If making has made us who we are, learning and teaching are inexplicably linked to developing the ability to make. That is to say "show and tell" goes back a long, long way.

Today's makers, from novice to advanced, have learned how to make using YouTube, a vast visual library of skills, techniques, and knowledge. We might take it for granted that this resource is so broadly available and so widely used but it has become indispensable. YouTube's ordinary-seeming how-to videos — often humble, raw, and personal — usually feature someone talking about what they do and showing you how to do it. We might think of it as the form of a very old story that we are wired to learn from.

In this issue, we celebrate YouTube's content creators, who not only are capable as makers but as communicators. They are using the advanced communications tools to do this most basic thing: sharing how to do something. What makes these videos so good is not the production values but the authenticity of the person sharing what they know and what they can do

For many, YouTube provides a way to learn those things that are not taught in school. It has also been a readily available substitute for learning in person. How many people do you know who could teach you flintknapping? You can find plenty of those people on YouTube.

During this pandemic, we are all making use of video learning, whether live or recorded. I've been learning how to mill flour for sourdough bread but also how to get rid of gophers, make masks, and prune sprawling tomato plants. I watch the videos differently than a TV show or movie. The end is not the video; it's accomplishing what I want to do.

YouTube is where there are so many who can teach and so many who want to learn. YouTube is like a campfire that we gather around in small groups, watching what each of us can do and learning to do it for ourselves. ◉

Maker Faire Comes Home

Ken Murphy's installation photographed by Mark Madeo

OUR SIGNATURE EVENT, BROUGHT TO YOUR LIVING ROOM

Virtually Maker Faire was a global, 24-hour online event in May, consisting of live video sessions and curated maker projects. See page 22 of this issue and watch session highlights at makerfaire.com.

I have had Virtually Maker Faire on all day playing in sync on three different monitors! Forgot how much Maker Faire feels like the maker equivalent of a Super-Bowl-nerdy-convention. So many interesting talks and interactive discussions. I'm really liking the online platform because I can take better notes. Thank you for making this happen!
—*Liz Chavarria, Atlanta, Georgia, via Twitter*

This new digital Maker Faire is an awesome idea. I have always wanted to attend one, but traveling makes it unrealistic for me. Thanks Make:!
—*Kevin Hanes, via makeprojects.com*
I thought the same thing. And no matter how many people are in a session, there's always space!
—*Stu Schwadron, via makeprojects.com*

A heartfelt note of appreciation for the Virtually Maker Faire event. It was a gamble to put on something that would approximate the excitement and energy of creativity that physical Maker Faires offer. The first two Maker Faires I attended left me elated and feeling like magic still existed in the world. And that's exactly what I felt after Virtually Maker Faire. Thanks to all the staff of *Make:* who clearly put in an incredible effort to rock the event! I'm so impressed with the quality of execution; it was a joy to watch it happen.
—*Dr. Matthew Wettergreen, Rice University, Houston, Texas*

Virtually Maker Faire helped with the sense of connection across the world, especially the rolling 24-hour nature of it — physical Maker Faires tend to be localised to one country.
—*Tanya Fish, VMF presenter, England, UK*

I am Camila and together with my brother Diego, we participated in Virtually Maker Faire as MoonMakers. We are a community makerspace and Mexican YouTube channel. Thank you — we loved being part of this incredible experience from home with our project and workshop. For us as young people it is very important, in these moments that are lived around the world, to be able to inspire people towards the maker movement just as you have inspired us.

—*Camila Luna, Apan, Hidalgo, Mexico*

MAKING CHANGE

Thank you for letting us makers share our knowledge. Your organization has changed so many things in my life! Thanks to you guys I started my own robotics company. You have been such a source of inspiration for me. I owe you big.

—*Pablo Martinez Díaz, Madrid area, Spain*

FARADAY FAN

"Solar Flares and EMP" (*Make:* Volume 72, page 102) is a genuinely interesting article and a fun weekend project. I think it can benefit everyone, since we rely on technology nowadays. I plan on making the protective sleeve at some point. The history part of it is very good too.

—*Ansel, 13, Oakland, California*

Tasty Tweets:

47ness ✪
@47ness

Fanart of @Odd_Jayy a real-life #steampunk gadgeteer; if you don't already follow his work YOU SHOULD 🤖

#somethingpositive #featurefriday

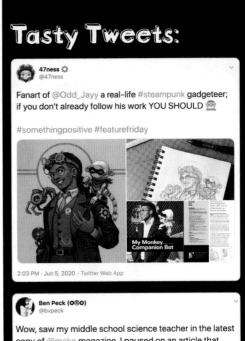

2:03 PM · Jun 5, 2020 · Twitter Web App

Ben Peck (O⊗O)
@bvpeck

Wow, saw my middle school science teacher in the latest copy of @make magazine. I paused on an article that caught my eye and bam, his photo! Shout out to Mr. Stith. Also, sorry I did so bad on your "memorize the table of elements song" challenge-- I've carried that for a while.

4:33 PM · May 28, 2020 · Twitter Web App

PROJECTS: Amateur Scientist

Solar Flares and EMP

Written by Forrest M. Mims III

102 make.co

Prepare for solar eruptions and nuclear weapons with a DIY Faraday pouch to shield electronic gear

Our dependence on electricity has made the entire world vulnerable to an existential threat. That threat is what might occur should there be a prolonged cutoff of electrical power to cities and even countries.

At least two kinds of events might cause a widespread loss of electrical power: a major coronal mass ejection (CME) from the sun directed toward Earth and the electromagnetic pulse (EMP) from the detonation of a nuclear explosion above the atmosphere. These events have the potential to shut down regional or even international electrical grids for days, months, or even years if high-voltage transformers at power plants and substations are destroyed by the huge electrical currents that a CME or EMP event can induce in high-voltage transmission lines.

The EMP from a nuclear explosion includes a secondary threat, for its nanosecond rise-time has the potential of damaging many kinds of electronic devices not connected to the grid.

SOLAR FLARES

On September 1, 1859, British astronomer Richard Carrington was sketching a cluster of sunspots viewed through his solar telescope when he saw what he described as "two patches of intensely bright and white light." Fellow astronomer Richard Hodgson also observed the phenomenon.

They had observed a massive solar flare accompanied by a CME that propelled an immensely powerful burst of energy directly toward Earth [Figure Ⓐ]. Less than a day later, the CME arrived and triggered brilliant, worldwide auroras. Telegraph operators reported large sparks in their equipment, and some received electrical shocks.

TIME REQUIRED:
30–60 Minutes

DIFFICULTY:
Easy

COST:
$25–$50

MATERIALS
» Conductive fabric intended for EM shielding, I used Mission Darkness fabric from mosequipment.com.
» Conductive tape such as TitanRF Faraday tape, also from mosequipment.com
» Static protection foil bag such as SCS Dri-Shield 3400, available at digikey.com
» Velcro tape, self-adhesive

TOOLS
» Scissors

FORREST M. MIMS III
an amateur scientist and Rolex Award winner, was named by *Discover* magazine as one of the "50 Best Brains in Science." His books have sold more than 7 million copies. forrestmims.org

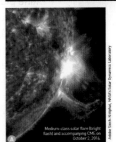

Medium-class solar flare (bright flash) and accompanying CME on October 2, 2014.

Ⓐ

Make: Community 103

MADE ON EARTH

Backyard builds from around the globe

Know a project that would be perfect for Made on Earth?
Let us know: *editor@makezine.com*

SYMPHONIC STEEL

MAKEZINE.COM/GO/SYMPHONIC-STEEL

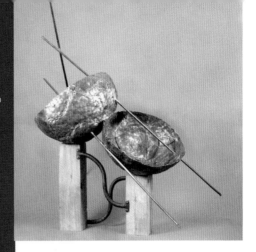

Seven steel appendages stretch upward, each gently twisted in space. With the press of a button, they begin to rotate, their curls turning into undulations that lend to the sculpture's name: *Corps de Ballet* (vimeo.com/415299403) — the term used for the group in a dance company that perform behind the soloist.

The piece is the latest by artist **Howard Sandroff**, who, at age 70, is 10 years into creating metal art after four decades of fame as a music composer and professor. His gradual switch in artistic mediums came when Sandroff suddenly found himself unable to compose music after a surgical accident; he enrolled in a sculpture and welding class to help rehabilitate himself. The medium brought a deep satisfaction: "I had no aspirations with sculpture," he says. "Just a love of it."

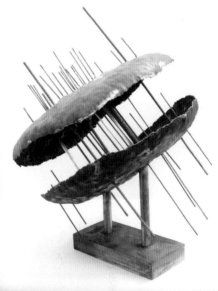

Over time, with his pieces overtaking their Chicago-area home, Sandroff's wife Mary Ann found a local business that could display one of his works. Other creations were entered into exhibits. A few started to sell. Sandroff gradually retired from his music commitments, and now spends all his time crafting metal sculptures and building various tools for his workshop from scratch. "Art is 10 percent technique and 90 percent magic," he says, exhibiting his humility about his success.

Most of his works are static sculptures that display the raw edges of a plasma torch and dimpling of a blacksmith's hammer, along with a patina that adds a depth of muted color. For his new kinetic piece, however, he refers to a realization that he's "always wanted my music to stand still, but for my sculptures to have motion." To generate that, he's incorporated a 12V motor and Arduino controller, along with a snaking drive chain and multiple gear cogs. He says it took him some time to create — "I haven't thought about angular velocity since 1967."

The piece is relaxing to watch, and Sandroff's enthusiasm about his endeavors is infectious. He sums it up nicely: "This is the best time of my life because I wake up every morning and make art."

—Mike Senese

Howard Sandroff

CIRCUIT BIKE LEONARDMOTORWORKS.COM

Taking the long, drawn-out geometry of 1920s-era German motorcycles as his inspiration, **Hunter Leonard**, age 21, is at the starting line of his custom-chopper endeavors. The Rochester Institute of Technology mechanical engineering student, currently sheltering at home in Wayland, Massachusetts, developed his passion for riding and repairing bikes during summer visits to his great uncle Bert, eventually hot rodding his own motorized bicycle for roaring drives around his neighborhood. Since gaining access to the workshop at his college, he's scratch-built four motorcycles and fixed and restored many others.

Leonard's latest creation, the Starrettania, is based on a 1928 Windhoff, and sports the classic big fenders and spring-loaded seat, along with a bolted-in-place subframe with chrome plated tubes. But where you'd expect to find a 4-cylinder engine is instead a grid of 16 Nissan Leaf batteries, along with four bright orange cables

leading to a rear-mounted 40HP hub motor.

To mount the classic and modern pieces together, Leonard designed a custom frame in SolidWorks and cut it on a Crossfire CNC plasma table. "When most bike builders start something new, they generally start with a frame that came off of another bike and make their modifications to that," he says. "No frame that I know of could replicate the look I was going for without requiring a plethora of modifications, to the point where it would be easier to start with nothing."

Still a work in progress, Leonard says the bike should get up to 90mph and has a predicted range of 80 miles. He's shown it at Maker Faires, and aims to log 5,000 trouble-free miles before embarking on the next build. So far, he's enjoying the ride: "There's no noise, no clutch, no gears, just twist the throttle and you're moving. It can be a very weird sensation to not hear very much when moving so fast." —*Mike Senese*

Make: Projects

A growing collaboration platform of makers, creators, mentors, and doers.

www.makeprojects.com

THE DEVICE DISSECTOR

MAKEZINE.COM/GO/DEVICE-DISSECTOR

Almost every modern device today is covered in a contoured, opaque plastic shell, designed for ergonomics and brand recognizability while obscuring the circuitry, motors, gears, and other mechanisms that make it work. Hong Kong mechanical engineer **Dennis Chung Yeung Poon** is changing that with clear-case conversions that give an instant look at the incredible scenery inside his favorite gadgets, from iPhones to GoPros to DJI drones.

His process, documented on the new YouTube channel **Useless Mod**, isn't for the faint of heart; to get these clear cases, Dennis creates them from scratch, dissecting an item entirely to extract just the external plastic shell. He creates a silicone mold of the plastic pieces, adding precise injection points, and uses clear epoxy

to cast a near-perfect match. He then carefully reassembles the device, and if all goes well, everything powers back up. But, spoiler — it doesn't always turn back on, and then Dennis has to troubleshoot to fix the issue.

"I want the videos to have a more educational purpose," Dennis says, "so people can learn from them rather than for them to only look cool."

One of the most effective of these so far is his impact driver rebuild, which clearly shows all the pieces (including snap-in-place battery), offering a good look at how the hammering aspect of these tools works. It also highlights his mold making technique, and some of the advanced equipment he has access to for making his cases.
—*Mike Senese*

Dennis Chung Yeung Poon

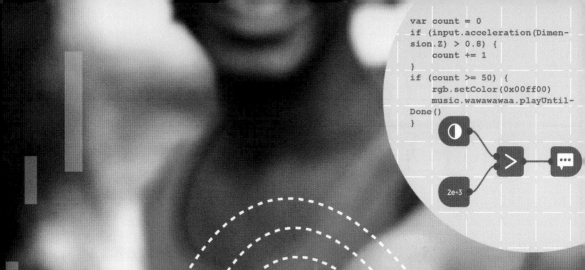

Reinventing a Material World

Written by Dr. Saul Griffith

WE NEED NEW TRICKS — **QUICK** — TO DECARBONIZE OUR STUFF AND THE WAYS WE MANUFACTURE IT

DR. SAUL GRIFFITH is founder and principal scientist at Otherlab, an independent R&D lab, where he focuses on engineering solutions for a clean energy, net-zero carbon economy. Occasionally making some pretty cool robots too. Saul got his PhD from MIT, and is a founder or co-founder of makanipower.com, sunfolding.com, voluteinc.com, treau.cool, departmentof.energy, materialcomforts.com, howtoons.com, and more. Saul was named a MacArthur Fellow in 2007.

Adobe Stock - dmitrymoi

Stand with the Children

This essay concludes our series on how makers can decarbonize our world. Read the whole series and find hands-on projects to make a difference at makezine.com/go/fix-our-planet. Adapted from *Rewiring America: A Handbook for the Climate Fight*, rewiringamerica.org/handbook. Also coming in 2021: an expansion of these ideas in *Electrify!* (MIT Press).

REWIRING AMERICA

SAUL GRIFFITH

Solving climate change isn't much good if we suffocate our oceans with plastics, kill our bees with pesticides, and pollute our waterways with excess fertilizers and toxins. Yes, we can decarbonize our energy supply with existing technology, but we've got to fix our industrial ecosystem too. Industry is where our climate change challenges collide with all our other environmental problems. The upside: There are huge opportunities for double wins — where we fix not only climate but also the other ills of the way we make the things in our lives.

Makers and engineers, the world needs you to fix manufacturing — to innovate and invent new, low-carbon methods and materials. And quick.

Our #1 Energy User

Our industrial economy is actually America's largest consumer of energy (≈32%) and a huge emitter of CO_2 and other climate warming gases. We can see the energy flow breakdowns in Figure **A** on the following page. The industrial sector, as defined by the U.S. agencies who measure our energy uses and carbon emissions, includes mining, construction, agriculture, and the biggest component, manufacturing.

This Sankey diagram, which is largely built on the semi-annual Manufacturing and Energy Consumption Survey (MECS), tells us quite a lot. You can dive deep down this rabbit hole. One thing that sticks out is just how much of our industrial sector is involved in finding, mining, and refining fossil fuels. Nearly 2 quads (quadrillion BTUs) are used in natural gas and oil exploration. Around 3 quads are used in refining crude oil into gasoline.

All told, something like 6 quads of the industrial sector's 32 total quads would disappear in a renewables-powered world! About 1 of those fossil quads is used just for making fertilizer. Fertilizer is good, and we need it, but we don't use it very efficiently and could use much less while maintaining a healthier and better food system.

While we can easily understand why our cars and homes produce CO_2 emissions from the giant amounts of fossil fuels we feed them, there's a lot more to understanding how the things that we purchase as "goods" contribute to emissions.

First, let's look at the flow of materials through the economy; it's an excellent complement to understanding the flow of energy. Figure **B** on the following page shows just how much stuff we move around. The 6,544 million metric tons of stuff we take from the natural world each year (in the U.S.) amounts to 20 metric tons per person!

(Funnily enough, this is without even counting CO_2. When we burn those 1,936 million metric tons of fossil fuels, they mix with oxygen to create CO_2 — around 6,700 million metric tons of the gaseous stuff, *more than everything else we push around combined!* Contemplate that before you get too enamored with the propaganda around carbon sequestration — you'd have to bury more CO_2, every year, than all the other stuff we dig out of the ground and take from forest and field. That's going to be one hell of an environmentally destructive process.)

On the bright side, looking at these giant material flows gives us the opportunity to contemplate a more sane version of carbon sequestration. Looking at those flows, especially the bigger ones,

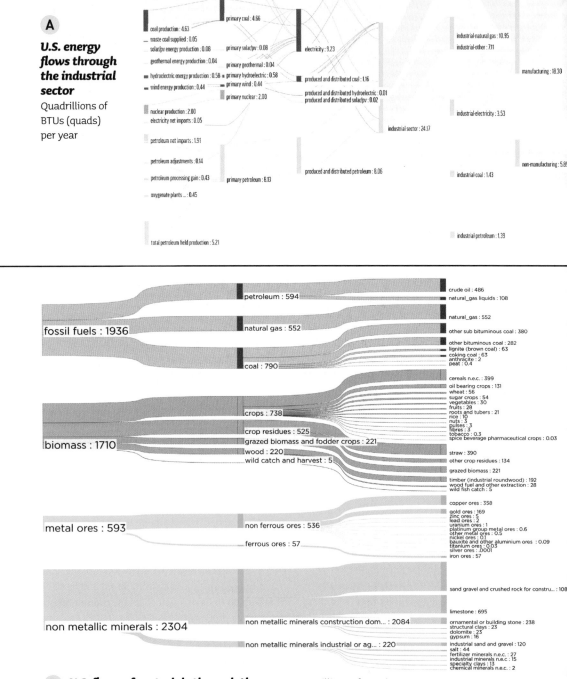

A

U.S. energy flows through the industrial sector

Quadrillions of BTUs (quads) per year

total biomass energy production : 2.51 — primary biomass : 2.51
natural gas net imports : 0.40
produced and distributed biomass : 2.38
waste : 6.01 — waste : 6.01
industrial electricity loss : 6.01

natural gas production (dry) : 11.51 — primary natural gas : 11.70
supplemental gaseous fuels : 0.03
produced and distributed natural gas : 9.31

primary coal : 4.66
coal production : 4.63
waste coal supplied : 0.05
solar/pv energy production : 0.08 — primary solar/pv : 0.08
geothermal energy production : 0.04 — primary geothermal : 0.04
hydroelectric energy production : 0.58 — primary hydroelectric : 0.58
wind energy production : 0.44 — primary wind : 0.44
primary nuclear : 2.00

electricity : 9.23

industrial-natural gas : 10.95
industrial-other : 7.11
manufacturing : 18.30

produced and distributed coal : 1.16

produced and distributed hydroelectric : 0.01
produced and distributed solar/pv : 0.02

nuclear production : 2.00
electricity net imports : 0.05

industrial sector : 24.17

industrial-electricity : 3.53

petroleum net imports : 1.91

petroleum adjustments : 0.14
petroleum processing gain : 0.43 — primary petroleum : 8.13
oxygenate plants ... : 0.45

produced and distributed petroleum : 8.06

industrial-coal : 1.43

non-manufacturing : 5.89

total petroleum field production : 5.21

industrial-petroleum : 1.39

---

petroleum : 594
crude oil : 486
natural_gas liquids : 108

fossil fuels : 1936
natural gas : 552
natural_gas : 552
other sub bituminous coal : 380

coal : 790
other bituminous coal : 282
lignite (brown coal) : 63
coking coal : 63
anthracite : 2
peat : 0.4

cereals n.e.c. : 399
oil bearing crops : 131
wheat : 56
sugar crops : 54
vegetables : 30
fruits : 28
crops : 738
roots and tubers : 21
rice : 10
nuts : 5
pulses : 3
fibres : 2
crop residues : 525
tobacco : 0.3
grazed biomass and fodder crops : 221
spice beverage pharmaceutical crops : 0.03
biomass : 1710
wood : 220
straw : 390
wild catch and harvest : 5
other crop residues : 134

grazed biomass : 221

timber (industrial roundwood) : 192
wood fuel and other extraction : 28
wild fish catch : 5

copper ores : 358

gold ores : 169
zinc ores : 5
lead ores : 2
metal ores : 593
non ferrous ores : 536
uranium ores : 1
platinum group metal ores : 0.6
other metal ores : 0.5
nickel ores : 0.1
bauxite and other aluminium ores : 0.09
titanium ores : 0.03
ferrous ores : 57
silver ores : .0001
iron ores : 57

sand gravel and crushed rock for constru... : 108...

limestone : 695

non metallic minerals : 2304
non metallic minerals construction dom... : 2084
ornamental or building stone : 238
structural clays : 23
dolomite : 23
gypsum : 16
non metallic minerals industrial or ag... : 220
industrial sand and gravel : 120
salt : 44
fertilizer minerals n.e.c. : 27
industrial minerals n.e.c : 15
specialty clays : 13
chemical minerals n.e.c. : 2

B *U.S. flows of materials through the economy* Millions of metric tons per year

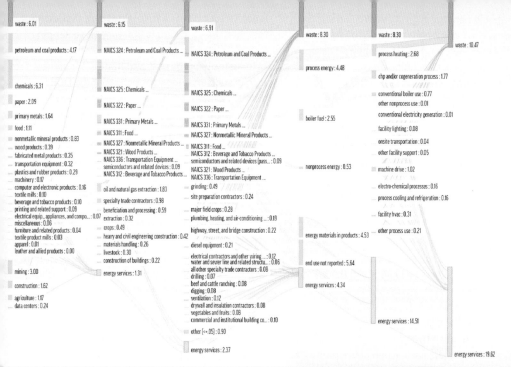

ask yourself: "Can I conceivably bury or sequester carbon in that flow?" Makers, if you can figure out how to answer yes, you'll make an enormous contribution to addressing climate change.

If we are to sequester carbon, it most likely will be by utilizing the large material flows we already engage in. Stash it in the soil and rock we move, or in forestry and wood products, or in concrete and drywall. It may not be as glamorous as carbon "air capture" but it's more likely and more reasonable. It is realistically slower than the pathways that've been modelled into UN IPCC emission reduction scenarios. That means you'll need to figure it out quick and get going.

So what I'm trying to do in this article is really two things. I'd like to show you the efficiency wins and technology transformations that can sharply reduce industrial energy flows, but I also want you to keep an eye open for opportunities to sequester CO_2 in the materials of our existence.

Embodied energy:
Thinking about the energy in stuff
Engineers think about the energy or carbon footprint of a product in terms of its *embodied energy* or its *embodied carbon* (see Table **C**).

C **Approximate embodied energies and embodied carbon for an array of common materials**

Material	Energy (MJ/kg)	Carbon (kgCO₂/kg)
Concrete	1.11	0.159
Steel	20.1	1.37
Stainless steel	56.7	6.15
Timber	8.5	0.46
Glue laminated timber	12	0.87
Glass fiber insulation	28	1.35
Aluminum	155	8.24
Asphalt (bitumen)	51	0.4
Plywood	15	1.07
Glass	15	0.85
PVC	77.2	2.41
Copper	42	2.6
Lead	25.21	1.57

Otherlab

Embodied energy is pretty easy to understand, which is why I personally use it as the reference number. As you can imagine, embodied carbon could vary greatly depending on the energy source that was used to produce the material. If we were making all of these materials with zero carbon electricity, most would have near-zero embodied carbon.

But this assumes the material is only used once. In reality, to compare all these things we need to recognize that the energy or carbon impact of an object is determined by the equation:

$$\text{Energy utility of a thing} = \frac{\text{Weight of thing} \times \text{embodied energy}}{\text{Lifetime or number of uses}}$$

This equation tells you some really important things. You can lower the weight of a thing — the strategy used by many companies to say, shave a few grams of plastic off your toothbrush — but the weight savings are generally really small (though often much hyped) and often designers use exotic materials like carbon fiber composites to achieve those weight savings, at the expense of the embodied energy actually going up! The other strategy lots of "green" companies use is material substitution — like all the "green" products made of bamboo. People associate bamboo with "green" but it isn't always so.

Oddly, it's that number underneath — the denominator — that makes the big, big difference: the number of uses, or the lifetime. If your bamboo toothbrush is only ever used once, it's an awful idea. If you use your carbon fiber bicycle for 15 years and 60,000 miles, it was an excellent choice. I've always thought about this as a giant opportunity. Think about how to make heirlooms: great products that people use for a long time, and amortize their embodied energy over a much longer period. (See *Make:* Volume 10, "Makers vs. Shakers.")

Vehicles are one category where technocrats obsess about embodied energy, and for good reason. Approximately half the carbon emissions of a typical car are in the production stage,

before it ever drives a mile. One thing that excites me about electric cars is that they're so simple they should last much longer.

Industrial energy use, and material resource use, is such an important topic that the DOE publishes fantastic studies on just how good could we get at producing various industrial materials. These are known as Energy Bandwidth Studies[1]. They're worth looking at, to see the big energy consumers and carbon emitters, and to find out where inventors and engineers can make the greatest impact to fix our planet. Let's look at a few:

Steel

The carbon emissions of steel are a result of the energy used in heating the steel and processing it, *and* of the coal used in the process of making the raw iron in the first place. All steels have significant quantities of carbon in them: "low–carbon steels" are ductile and pretty strong, "high–carbon steels" are more brittle but super strong. Today 69% of steel is recycled in the U.S. **FIX IT:** Today the heat for the steelmaking process in most places comes from natural gas, but there's no reason it can't come from clean electricity. And there are companies all over the world working on ways to add the carbon content without having to add it as coal in the blast furnace. A Rearden Metal-sized fortune[2] will go to whoever succeeds.

Concrete

I've always been astounded with the statistics on concrete. In the U.S. we produce almost 2 tons per person per year! The stuff is now everywhere. Joni Mitchell was bang on target when she sang of paving over paradise. It's estimated that 8% of global emissions come from cement alone. Half of those emissions are from the energy required, the other half are emitted in the creation of the *clinker* — the lime-based binder that holds it all together.
FIX IT: Limestone ($CaCO_3$) is heated to become lime (CaO) which leaves a CO_2 leftover. But it doesn't have to be

this way. We should be able to make cement that *absorbs* CO_2 through its lifetime. And we certainly should be able to build with less concrete. Covering ground with concrete has negative effects on drainage, soils, and more. I'm sure we can do better.

Aluminum

Most aluminum is already made with electricity, so once again, in theory we can make it without carbon emissions, but the energy input is not the only source of carbon. In the arc furnaces that we smelt aluminum in today the electrodes are carbon. This is the source of much of their carbon emissions. Today 69% of aluminum is recycled in the U.S.

FIX IT: Apple recently worked with Alcoa and Rio Tinto to create the first batch of carbon–free aluminum. I've personally always found aluminum to be a wonder material, so I'm glad we are on a good track for carbon-free Al.

Paper

In concept, paper can be a zero net carbon product. About 63% of paper and paperboard are recycled in the U.S.

FIX IT: The huge amount of energy used in the paper and pulp industry (more than 2 quads!) is mostly used in separating the cellulose fibers from the lignin glue that holds trees together. We can invent something better.

Timber

Wood is good. I like to think it's the second best method for carbon sequestration, after books!

FIX IT: We need to do much better forestry management, but people are already building wooden multi-story housing and wood is really a perfect sustainable building material. I once planted 30,000 trees. They all grew up. They could have been my entire lifetime's supply of construction materials. And then some.

Glass

Glass can be recycled basically infinitely, but it does use a lot of energy to produce it (principally because the melting point of glasses is so high). Today 34% of glass is recycled in the U.S.

FIX IT: We are getting good at making stronger, thinner, tougher, glasses, but maybe all we really need is a cultural shift back towards reusable glass packaging. It's cleaner and chemically much safer than storing your food in plastic.

Plastics

Plastics on their own, unless we change things quickly, emit 10%–13% of our remaining carbon budget[3]. This isn't as obvious as you think. Plastic molecules have big long carbon backbones and last forever, so you'd think we could sequester carbon in them. (Drill for oil to mold giant plastic dinosaurs that we could re-bury to sequester carbon! Kidding.[4]) What actually happens is that in the creation of *olefins* — the precursor to most plastics — there are large amounts of nitrous oxide emissions, and they are even worse for warming than CO_2. Less than 10% of plastic is recycled in the U.S.

FIX IT: Recycling won't work. We need an entirely new pathway to plastics, but even if we achieve it, they'll still gather in the oceans. Because of all this, I think we should use paper and glass and metal and more reusable containers, but also we should invest heavily in synthetic biology pathways to a new kind of polymer that would quickly biodegrade the way leaves do. Leaves don't end up as ocean microplastics.

The Future We Make

The upshot: If we are wise, our moonshot science investment should be in studying and inventing materials systems, especially polymers, that don't degrade our environment, or use excess energy.

And these sectors are just the beginning. Look again at our industrial energy consumption (Figure A) by sub-sector. Can you imagine a better way to do any of these things (just as examples): grinding of materials (0.49 quads), electrochemical processing (0.16 quads), or food processing (1.11 quads)? Do it, and you could be a captain of that new industry. ✏

• • •

[1] https://www.energy.gov/eere/amo/energy-analysis-data-and-reports
[2] Please excuse my Ayn Rand *Atlas Shrugged* reference.
[3] https://www.ciel.org/wp-content/uploads/2019/05/Plastic-and-Climate-FINAL-2019.pdf
[4] Come to think of it this is what most people's carbon sequestration plans are anyway!

PROTESTS MATTER

Written by Keith Hammond

50+ HOW-TO'S FOR FLEXING YOUR CONSTITUTIONAL RIGHTS

Taking to the streets in peaceful protest is a sacred American right, enshrined in the First Amendment to the Constitution. And it's sure not any easier during a pandemic.

Here's how to make what you need for safe, peaceful, and effective protest. This list is just a start — find these and many more at makezine. com/go/make-change. Stay healthy out there!

SIGNS, BANNERS, EFFIGIES, PERFORMANCE

Creative Direct Action Visuals — Excellent guide to banners and flags, stencils, posters, T-shirts, puppets, etc., from Ruckus Society: ruckus.org

Making signs and posters — New York City and other jurisdictions don't allow wooden sticks or metal in signs; know your local law.

- Collapsible sign — by Paul Spinrad, makezine. com/projects/collapsible-demonstration-sign
- LED-based "neon" sign — by John Park, learn.adafruit.com/led-neon-signs
- LED matrix protest sign — by John Park, learn.adafruit.com/led-protest-sign
- Protest signs generally — by Felicity Shoulders, medium.com/@faerye

Free graphics for protest — from Just Seeds, justseeds.org/graphics

Banner hoisting, banner drops, effigies, giant inflatables — destructables.org

Tactical Performance, Electoral Guerilla Theater — books by L.M. (Larry) Bogad of Clandestine Insurgent Rebel Clown Army, routledge.com

PERSONAL PROTECTION

Pack a Wellness Bag — Shatter-resistant eye protection (don't wear contacts), shoes you can run in, change of clothes and vinegar-soaked bandana (for chemical exposure), more. From

Black Lives Matter: makezine.com/go/blm-bag

COVID-19 Go Kit — Pandemic addendum: face masks, hand sanitizer, gloves, and more. From Blackout Collective and Movement for Black Lives (M4BL): makezine.com/go/m4bl-kit

Medic's Kit — First aid for fellow protestors, from Frontline Wellness United: frontlinewellness.org

Remedies for pepper spray / tear gas —
- Saline solution or antacid/water mix, frontlinewellness.org
- *Illustrated Medical Pamphlet for Protests*, from Occupy Oakland: destructables.org

Tear gas traffic cones — Hong Kong tactic: cover a tear gas grenade with a traffic cone, pour water in to neutralize it: youtu.be/hpqEQARnVbs

STREET ART AND POSTERING

How to Paint a Street Mural — by David Solnit: makezine.com/go/solnit-street

Street mural materials and logistics — Tips from Black Lives Matter mural team in DC: makezine.com/2020/06/19/how-to-paint-street

Chalk drawing and layout — chalkupy.org/how-to

Stealthy sidewalk stencil box, wheat paste for postering, billboard "improvement," asphalt mosaics, DIY newspaper headlines — destructables.org

ORGANIZING AND OCCUPYING

Action Strategy: A How-to Guide — from Ruckus Society: ruckus.org

Apps to use to organize protests — vox.com/recode/2020/6/3/21278558

Protest and campaign tactics — beautifulrising.org and beautifultrouble.org

Lockboxes — for human-chain blockades and sit-ins: destructables.org/node/59

Topple a statue safely — from *Popular Mechanics*: popularmechanics.com/science/a32870657

SECURITY AND PRIVACY

Security Culture for Activists — from ruckus.org

Defeat phone surveillance — Leave your phone home; opponents can use it to spy on you. Or:
- Make a Faraday pouch — to block phone signals. From Forrest Mims in *Make:* makezine.com/projects/amateur-scientist-solar-flares
- Disable facial/thumbprint authentication — Use password/PIN instead: wired.com/story/how-to-protest-safely-surveillance-digital-privacy
- Digital privacy and security measures for staying safe while protesting — by Violet Blue: blog.adafruit.com/2020/06/05
- Tools for blurring faces and anonymizing your photos — techcrunch.com/2020/06/06/protesters-blur-faces

Defeat face recognition and surveillance cams —
- Confuse the cameras with weird face makeup, "adversarial" images, infrared LEDs, or a good old hoodie, bandana, and sunglasses. From *Make:* makezine.com/2020/06/16/face-jam
- More tips and tricks: wired.co.uk/article/avoid-facial-recognition-software

INTERACTIONS WITH POLICE

Know your rights — aclu.org/know-your-rights/protesters-rights

Your local number for National Lawyers Guild — Write it on your body in Sharpie. They can help to get you out of jail. nlg.org/massdefenseprogram

Download an SOS phone app — Alerts friends and family you're being arrested.

Siri shortcut: Pulled Over By Police — Quiets your phone, records video, texts a friend you've been pulled over, with location, then sends the video to your friend. By Robert Petersen: reddit.com/r/shortcuts/comments/9huqiw

RADIO AND SITUATIONAL AWARENESS

Using a police scanner — Perfectly legal in most places. Programming in trunked radio systems: youtube.com/watch?v=yWrOAEaTRv0

DIY police scanner (analog) — Use your computer as a scanner, with a $20 software-defined radio (SDR) dongle: webcommand.net/index.php/2018/01/19/build-police-scanner-20

DIY police scanner (digital + analog) — Your PD uses digital audio? Add digital audio decoder software to your project. Links for MacOS, Windows, Android, and Raspberry Pi builds: makezine.com/go/make-change

Pirate radio — Broadcast your own FM station:
- Using a Raspberry Pi, from *Make:* makezine.com/projects/raspberry-pirate-radio
- *The Complete Manual of Pirate Radio* by Zeke Teflon: destructables.org ◓

IT'S BACK!

VIRTUALLY MAKER FAIRE

Written by Craig Couden

MAKERS AROUND THE WORLD LOGGED ON FOR AN EPIC 24-HOUR, GLOBAL SHOW AND TELL

Featuring online video sessions and a curated collection of maker projects, many of which were developed in response to Covid-19, *Make:* hosted the first Virtually Maker Faire on May 23, 2020. Like previous Maker Faires, it was a showcase of what people are currently doing, what they're interested in, and what they want to share with others. But unlike physical Faires, this was worldwide, and you can still see many of the exhibits and talks right now!

Through an open Call for Makers, we hosted makers representing 25 countries who shared over 350 presentations, demonstrations, and online project exhibits concurrently on multiple channels. "This was a phenomenal event," said Dr. Sally Applin, in a post-event wrap-up with *Make:* founder Dale Dougherty. "I loved dipping into different sessions as much as I could in the past 24 hours." Over 110 sessions are archived at makezine.com/go/vmf-playlist. Some highlights:

EEPYBIRD: UNLOCKING CREATIVITY

EepyBird, aka the Diet Coke and Mentos guys, kicked off Virtually Maker Faire with a fun look inside their creative process — and how they're using it to generate new ideas for today's challenges. "What we do is come in and say, 'Today we're going to take things one step forward,'" explains EepyBird's Fritz Grobe. youtu.be/OY60jUVyDU4

PERFORMERS RESPONDING TO STAY-AT-HOME

Touring magician (and Maker Camp MC) Mario the Maker Magician and wife/partner Katie Marchese hosted a panel discussion with performers including Cressie Mae of Circus Workout Party and magicians Ran'D Shine and Kameron Messmer about the successes and challenges of adapting live performances in a world of stay-at-home orders. "We're being forced to change, and we're learning things that we'll absolutely bring into our career that we wouldn't have otherwise," said Katie. facebook.com/5873603189/videos/2906067226178324

HOW TO MAKE A PUPPET!

Puppeteer Adam Kreutinger gave a live demo of how to make your very own fluffy, goofy looking puppet that was full of tips and tricks. youtu.be/XbaeQTtueEE

MASKS FOR EVERYONE: SERVING UNDERSERVED COMMUNITIES

Hosted by MakerMask.org, this panel discussion looked at challenges and solutions around getting face masks to low-income individuals and communities, as well as making masks accessible for people with visual impairments and sensory challenges. "We got with our local groups that served our local communities already," said panelist Rachel Sadd, "and created partnerships for distribution." youtu.be/SEN7SwTFHsY

HOW TO BECOME AN INVENTOR

Estefannie from Estefannie Explains It All and Ruth Amos from Kids Invent Stuff talked about bringing ideas to life and encouraging creativity for young inventors and makers. "There's a lot of research about when young people form their ideas about careers. People thought it was a lot later, but actually it's primary school," said Amos. facebook.com/makemagazine/videos/2724172664484128

SOPHY WONG: STUDIO TOUR AND WORKS IN PROGRESS

Sophy Wong, designer of wearable tech, costumes, and more, took us on a virtual tour of her enviable home workshop and talked about how she's staying productive, creative, and inspired during shelter-in-place orders. "We do so many kinds of making, we have so many different materials ... it's just a sea of plastic bins," said Wong, showing one of her many intricately organized shop spaces. youtu.be/7aYi3lGs96s

DIY HEROES: MEET THE MAKERS FEATURED IN MAKE: VOL. 73

Meet some of the inspiring makers who are fighting Covid-19, and go behind the scenes of *Make:* magazine in this panel discussion featuring Executive Editor Mike Senese and Senior Editor Keith Hammond along with profile subjects Robert L. Read from Public Invention, Sabrina Paseman from FixTheMask.com, and Bill Hemphill from East Tennessee University. youtu.be/d2yF72J-uo0

ARCATTACK'S TESLA COIL MUSIC

To finish out Virtually Maker Faire, brothers Joe and John DiPrima, aka ArcAttack, live-streamed their unique Tesla coil-powered rock show from their shop in Austin, Texas. youtu.be/oc1z0s9Clo8

VIRTUALLY MAKER FAIRE PROJECT SHOWCASE

There's even more Virtually Maker Faire to explore at makeprojects.com. While you may have set your schedule to catch a live streaming session, perusing these projects is more like wandering through the Faire until something catches your eye. Whether it's a face mask that automatically closes when other people approach, cute robot companions, or DIY face shield and ventilator designs, there are plenty more projects to see. makeprojects.com/viewAll/?category=virtual-maker-faire

"Once again, the maker movement is on the edge of what to do next, which is go online, do what you need to do, work in a distributed way globally." —*Dr. Sally Applin, @anthropunk* ⊘

Virtually Maker Faire is produced with the support of Make: Community Members. Become a member today at make.co!

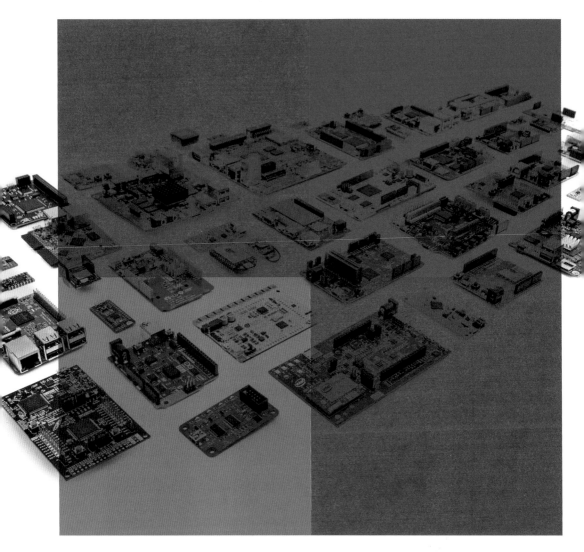

So many choices...

Written by Mike Senese

With more features and control than ever, dev boards give makers limitless options for their projects

There's no time like right now to make that thing you've always been wanting to build, and that's especially true if your project involves electronic control or some sort of computing need. The latest dev boards on the market are smaller, more powerful and feature-laden, and cheaper than ever. Meanwhile, your choices of software for programming them have expanded to include super-simple block-style and Python-based drag-and-drop methods across a broad range of board options. Whether you're building a single, permanent project, or creating proof-of-concept electronics that will turn into a custom PCB, there are many — even limitless — options to get you started.

So how do you pick? Size? Specs? Convenience? We've reached out to some of our high-tech pals to get the skinny on their personal processes for board selection.

Jorvon Moss (instagram.com/odd_jayy)

Since a lot of my projects are wearables, I often think of size and weight. I use **Adafruit's Trinket** for servo and LED control, **Pro Trinket** for more than two servo controls, then if I want internet access or online control I go for the **Raspberry Pi**. For everything else I run for an **Arduino Uno** or **Arduino Nano**. [See Odd Jayy's "My Monkey Companion Bot" in *Make:* Volume 73.]

Sophy Wong (sophywong.com)

I generally choose the board for my project based on the functions I'll need and the size requirements. I keep on my desk all the basic things I need to whip together a quick electronic prototype. It's easy to just grab a breadboard and start working out my circuit. I can develop my code at this point too, and transfer it to the final build when I'm ready. When I'm finished with my project, I disassemble the prototype and keep the components in my kit for future projects.

For small projects, my go-to board is the **Adafruit Gemma**, which is compact, easy to power, and has a convenient on/off switch. I'm also starting to play with making my own custom boards, which was a bit daunting at first, but feels really empowering as a designer. [See Sophy's "Cosmic Cosplay" in *Make:* Volume 69.]

Jorvon Moss

Sophy Wong

Hep Svadja, Tiffany Chien, Jorvon Moss, Sophy Wong

Alex Glow

Debra Ansell

Alex Glow (alexglow.com)

There are a few favorites I always reach for:

- **Adafruit's Circuit Playground Express**: LED ring, built-in speaker. Good for standalone interfaces and lights, like a DIY lightning cloud.
- **BBC micro:bit**: 5×5 LED matrix, 2-way radio, lots of programming options. Great for bike/skateboard lights.
- **Teensy**: A fast heavy lifter, with an audio breakout available. Good for motors and music, e.g., filter pedals.
- **TinyCircuits TinyLily:** Teeny-tiny, simple, washable. Good for wearables, like my "mind-altering gadgets."
- **Particle Photon**: Easy Wi-Fi integration (and beyond). Good for IoT without the hassle of ESP8266, including the brainwave-controlled light I built with Moheeb Zara.

[See Alex's "AI Robot Owl" in *Make:* Volume 66.]

Debra Ansell (geekmomprojects.com)

I'm always on the lookout for super small microcontrollers. If I see a new one that looks interesting, I'll buy it "just in case."

I really like prototyping in CircuitPython, so my go-to boards for small wearable projects are **Adafruit's Trinket M0** and **Gemma M0**, and I usually have a stash of them on hand. I also keep some ATtiny85 chips around for when I want to make something *really* small or when I test out a project that won't let me reuse the board afterwards, like embedding LED circuits in resin. For larger projects where I want sensors, Bluetooth or both, I'll prototype with a **Circuit Playground Express Bluefruit**.

In the past, my custom PCBs have mostly been accessories for an external microcontroller, but recently I've been creating self-contained PCBs with SMT LEDs and an ATtiny85 controller, powered by a CR2032 battery. [See Debra's "LED Inner-Glow Heart" in *Make:* Volume 71.]

Liz Clark (blitzcitydiy.com)

Lately most of my projects are written in CircuitPython, so that narrows things down to an M0, M4, or nRF52840 board if I'm using BLE.

I try to not use a board that's overkill for a project. For a simpler LED project I'll use a small board, like a **Trinket M0**, rather than something bigger like a **Metro M4**. I try to use a board that has as close to the number of pins that I need to try and keep things as compact as possible. This also reduces cost since board price is also usually related to size.

If I need battery power, I'll try to use a board that has an onboard JST plug for a LiPo, like the **Feather** series of boards. I'll also try and have as many features that I need for a project (Wi-Fi, a screen, speaker, capacitive touch, etc.) built into the board. That simplifies things for assembly and code since there will usually be a dedicated library for those peripherals on that board.

Kitty Yeung (artbyphysicistkittyyeung.com)

For prototyping, I usually start with an easy board like the **Arduino Uno** to check that the circuit and code work. For wearables and tech fashion design, the boards need to be well-hidden, so I look for the smallest capable board.

Then I look at how to power the board. It's best if they already have a LiPo battery port or coin cell holder — having a USB battery pack is a

Lady Red Beacham, Justin Ansell, Debra Ansell, Liz Clark, Kitty Yeung, Caroline Lough, Brian Lough

Liz Clark

Brian Lough

Kitty Yeung

Brian Lough (blough.ie)

The first thing I do is ask myself the question: "Is there any reason why I shouldn't use an **ESP8266**?" This is my go-to board with my favorite flavor being the **Wemos D1 Mini**-style boards. They are small, breadboard compatible, and can be programmed via USB. They also cost less than $3 delivered so I usually keep a healthy stock of them! It would be a struggle to incorporate its features into a design for less than the cost of one.

Battery needs are one of the main reasons I might use something else. For those cases, I usually select **Unexpected Maker's TinyPICO** board. The **ESP32** is also a good choice here. For lower powered projects such as badges I usually select something like an ATtiny13 chip, ideal to run off coin cell batteries.

USB HID (emulating a keyboard or mouse) is another feature I use regularly. For this: the **Arduino Pro Micro** (ATmega32U4) or the **ESP32-S2**. [See Brian's "Travel Light" in *Make:* Volume 59.]

For even more help in your dev board decision-making, consult our Guide to Boards, included with this issue. ❷

nuisance. An onboard power switch is convenient. For projects that need conductive thread sewn to the fabric, I need large pin holes that can feed a needle through, like on the **Adafruit Flora** and **SparkFun LilyPad** series.

I try to design all the above elements into a custom PCB if I can't find an existing board that satisfies all the needs. With this, I can also control the board footprint and appearance. [See Kitty's "Novel Nails" in *Make:* Volume 62.]

Python in physical
computing is on the rise

Play time:
Piano in the
Key of Lime, by
CircuitPython
developer
Kattni Rembor.

Python
on Written by Helen Leigh
Hardware

Helen Leigh is a hardware hacker who specializes in music technologies
and craft-based electronics. Say hello to her on Twitter @helenleigh.

Python is one of the fastest growing languages in the world. It's accessible enough for schools to teach as a first programming language but powerful enough to handle the complexities of some of the most widely used web services, including Instagram, Spotify, and Netflix. And now Python's star is rising in the world of physical computing too.

Languages like C and C++ and their cousin Arduino used to dominate hardware, from hobbyists in garages to engineers in industry. But these days, thousands of makers are switching to Python to control hardware, whether they're programming microcontrollers or designing microprocessors. Today more than 130 microcontroller boards support Python, including the original Pyboard, the blazing Teensy 4.0, the tiny Serpente, a host of Adafruit boards, and even five Arduinos.

In this article, we'll explore the reasons behind this shift, talk about why you might (or might not!) want to use Python in your next project, and take a look at some cool hardware projects and technologies that use Python.

Best for Beginners?

Beginner programmers often ask, "Which language is best to learn: Python or C++?" There's no definitive answer, of course, but here are some reasons why many programmers might well respond with "Python." Let's start off with **user friendliness**. When you upload your code to an Arduino board, your computer compiles it into a machine-readable binary format. The semicolons and curly braces in your code help the compiler to identify where statements and function blocks

begin and end. Python doesn't require semicolons to end lines and instead relies on indentation to distinguish code blocks. This has the benefit of making Python code very **human-readable** and clean.

Python is also not as nosy about your variables because as an interpreted language it doesn't require the programmer to manage memory — Python **manages memory for you**. Data types for variables in C languages (floats, integers, `const signed long`, etc.) have to be declared, but Python will happily interpret these for you. Because the language isn't compiled but interpreted, you can also skip all those pesky considerations that have to do with memory management, like "Should I use a constant? Is this a long or a short? Do I need to use a pointer?"

Python is also a fantastic choice for rapid prototyping because of its readability and ability to **run without needing to compile**. When you upload your code to an Arduino, it has to compile and flash the hardware, which takes time. When you save your CircuitPython code to a board, it skips that step and executes the code as soon as you hit Save. But hold up! This doesn't make Python code faster. Compiled languages like C++ are actually much faster, because they don't have to be interpreted at runtime. For projects where speed is of the essence (for example, when using extra demanding sensors or actuators), this is something to consider, but for most beginner maker projects it won't likely be an issue.

MicroPython and CircuitPython

Two Python implementations are heating up the microcontroller world. First up is **MicroPython** (micropython.org), created by Damien George for the STM32 chip but now popularly used by both the BBC micro:bit as well as the ESP32. MicroPython is a direct translation of Python with its own interpreter.

The other big player is **CircuitPython** (circuitpython.org), a fork of MicroPython maintained by Adafruit Industries. CircuitPython's goal is to lower the barrier to entry for beginner programmers, so the Adafruit team have decided to sacrifice some functionality in exchange for simplicity. We'll show off some of its features later in this article.

Kattni Rembor, Adafruit

Space saver: Mini SAM is a Lego minifigure-sized microcontroller board by Benjamin Shockley, based on the Python-compatible Microchip SAMD51G.

The BBC **micro:bit** has deployed nearly **5 million devices into education,** including 45,000 sent to schools across Croatia, and one sent to every single fourth grade child in Denmark.

Get schooled: Millions of students can learn Python on the BBC micro:bit.

Community, Education, Creativity

The other big advantage of using Python on hardware is the people that come with it. Lots of new users are coming over from the **Python community**, many of whom have little or no experience with hardware. For developers who've never touched a microcontroller, something like the Circuit Playground Express (CPX) allows them to experiment with the joy of DIY sensing systems, blinky lights, and buzzy music without needing all the tools, electrical engineering knowledge, and fabrication skills to put together a circuit.

There's also the cultural aspect: Python is well known as a language that has a comparatively **diverse, welcoming user base**. This is a tradition that Adafruit are continuing with their family-friendly, tightly moderated Discord channel which has over 17,000 members who you can ask for help troubleshooting your project.

Boards running Python are extremely popular **learning devices in education**, from schools and colleges to families who want to explore technology at home. The BBC micro:bit alone has deployed nearly 5 million devices into education, including 1 million in the UK, 45,000 to schools across Croatia, and one for every fourth grade child in Denmark. Microsoft's MakeCode, a popular visual block-based programming

environment for Minecraft, micro:bit, CPX, and other education-focused technologies, added Python support this year, so students can toggle back and forth between visual blocks and text-based Python while they learn the ropes.

Another significant group using Python on hardware are people who are looking to add technology into their existing practice — **artists, designers, scientists, and cosplayers** who use hardware as a tool to make their work easier, cooler, or more exciting. These people, like many Arduino users, don't necessarily want to learn to code to become a programmer, they want to learn to code so they can make their ideas come to life.

This wide variety of users are exactly what the team behind CircuitPython — including Scott Shawcroft, Kattni Rembor, and Anne Barela — keep in mind when they make decisions about the direction they take their work in. Kattni explains that this is why every board they work on comes with a wide range of learning guides and ready-to-use examples that are designed around creative applications of the tech. Scott says their approach is to minimize the number of computing concepts people need to know to make useful and fun things when they're starting out, saying that "If it ruins the **beginner experience** then we're not interested."

Bigger, Slower, Hungrier

Python on hardware is still a work in progress and won't be the best option for every project. Some common complaints are that it **requires a lot of memory** and uses up **a lot of power**. For example, one badge hacking team wanted to use CircuitPython for a convention badge but weren't able to, given that a badge needs to last for a weekend of hacking on two AA batteries.

As mentioned, Python is also **slower than C++** so if you're weighing up CircuitPython on the Circuit Playground Express versus C++ code on the eye-wateringly fast Teensy 4.0, then Python might not come out on top for your project.

There has always been room in our community for different languages that serve different purposes. It's clear that Python on hardware has a great deal to contribute and that it's here to stay.

Getting Started with CircuitPython

Let's get started with Python on hardware. We're going to use CircuitPython and Adafruit's Circuit Playground Express board (Figure Ⓐ).

1. SETUP

The easiest way to start coding Python on hardware is using the Mu code editor, a free and open source piece of software. Go to codewith.mu and install the package that works best for your system. Then plug in your board with a USB cable that supports both data and charging. If you see a drive called *CIRCUITPY* appear on your computer then you're good to go.

No *CIRCUITPY*? No worries! If you see a drive called *CPLAYBOOT* you'll need to set up your board. Follow the instructions at Adafuit's guide, makezine.com/go/cp-quickstart.

Go ahead and open Mu. If it doesn't detect a board the program will prompt you to select your mode. For the Circuit Playground Express you need to select Adafruit CircuitPython, but you can also choose modes for the BBC micro:bit, PyGame Zero, or Python 3. You're ready to code!

2. BLINK! YOUR FIRST CIRCUITPYTHON CODE

Blinking an LED is the hardware version of **Hello, World!** Let's test our Circuit Playground Express ("cpx" in code) by blinking an onboard

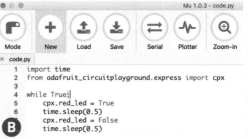

```
import time
from adafruit_circuitplayground.express import cpx

while True:
    cpx.red_led = True
    time.sleep(0.5)
    cpx.red_led = False
    time.sleep(0.5)
```

LED. Here's the blink code (Figure Ⓑ) to turn an LED on and off every second. Make sure to indent the lines of code underneath **while True:**.

The first two lines of this code are importing libraries: one to manage time and another with all the settings for your board. The main body of your code lives underneath **while True:** — your board will loop through all the things that are indented underneath this line. In this case, **cpx.red_led = True** tells our board to turn on the red LED, **time.sleep(0.5)** tells it to pause for half a second, **cpx.red_led = False** turns the red LED off, and one more **time.sleep(0.5)** adds in another pause. These pauses control how long the LED stays on and off.

Once you're ready to send the blink code to your board, save the file as *code.py* on your *CIRCUITPY* drive. Once it's done saving, take a look at your Circuit Playground Express and you should see a small red LED blinking away every second.

One of the cool things about coding hardware using CircuitPython is that if you want to change something in your code you can simply make the change and hit Save. Your new code will be automatically sent to the board and start running straight away. This makes for much faster coding cycles — great for trying out new things or quickly checking changes. Try it out by changing the timing of your blinking light. The Mu editor also has a code checker that will warn you if you

Micro:bit Educational Foundation, Adafruit

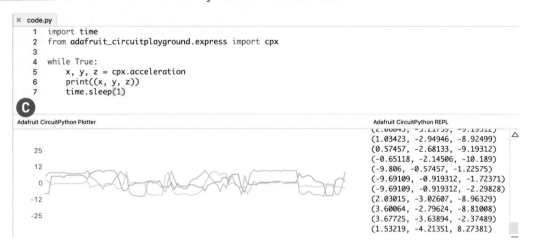

```
  code.py
  1    import time
  2    from adafruit_circuitplayground.express import cpx
  3
  4    while True:
  5        x, y, z = cpx.acceleration
  6        print((x, y, z))
  7        time.sleep(1)
```

Adafruit CircuitPython Plotter

25
12
0
-12
-25

Adafruit CircuitPython REPL
```
(2.00645, -5.21755, -9.19512)
(1.03423, -2.94946, -8.92499)
(0.57457, -2.68133, -9.19312)
(-0.65118, -2.14506, -10.189)
(-9.806, -0.57457, -1.22575)
(-9.69109, -0.919312, -1.72371)
(-9.69109, -0.919312, -2.29828)
(2.03015, -3.02607, -8.96329)
(3.60064, -2.79624, -8.81008)
(3.67725, -3.63894, -2.37489)
(1.53219, -4.21351, 8.27381)
```

have any obvious errors. Delete one of the tabs underneath **while True:** and click on the Check button to see what happens.

Working? Amazing! You're up and running. Problems? Don't worry! Head over to the Adafruit Learn guide at makezine.com/go/cp-quickstart and you'll find more detailed instructions and troubleshooting tips to get you going.

3. READING SENSORS

Like a lot of modern boards designed for education, the Circuit Playground Express has lots of cool sensors built into it. If you want to find out what they're sensing, you're going to want to get to know the Serial and Plotter features of Mu.

The serial console is how we receive information from our board, including sensor data and error messages that help us figure out what's wrong. To connect to the serial console, click the Serial button on the Mu toolbar. The plotter displays data from your board, with data on the vertical y-axis and time on the horizontal x-axis. Here's some code (Figure **C**) that shows us data from the accelerometer — a sensor onboard the Circuit Playground Express that detects motion in three different directions: x, y, and z.

Again, we're importing our libraries on lines 1 and 2, then setting up a **while True:** loop for our main block of code. Inside the loop, the line **x, y, z = cpx.acceleration** assigns accelerometer data values to the variables x, y, and z, while **print((x, y, z))** prints that data to our plotter (the wiggly lines on the bottom left) and to the

serial monitor (the numbers on the bottom right). Finally, **time.sleep(1)** pauses the board for one second before starting the loop again.

4. SUPER FAST CODING WITH REPL

One of the most exciting things about using Python on hardware is *REPL*, an interactive way of talking to your board that makes experimenting with code super speedy. REPL stands for *Read, Evaluate, Print, Loop*, which tells you how it works: It reads your Python commands, evaluates your code to figure out what you want, prints any results, then loops back to the start.

Let's try it out. Click the Serial button to open a serial connection to the connected device, then hit Ctrl-C. Press any key to enter the REPL. Once you're done, press Ctrl-D to exit the REPL.

Let's take a look at the example in Figure **D**. I entered the REPL by typing Ctrl-C, then pressing a random key. The board told me a bit of information about itself, then prompted me to enter a command with **>>>**. I imported the usual libraries, then I tried to take a look at the readings of the light sensor using the command **print("light:", cpz.light)**. However, I didn't get a reading. I got an error, telling me it didn't understand **cpz**. That's because I had a typo! REPL is great for sniffing out errors like this. Next, I tried to get a light sensor reading again, this time without the typo. To save me typing the whole command again, I used the Up key to select my previous command, then edited it to remove the typo. Success! Next, I used the Up key again to print another light sensor reading, this time with

```
× code.py
    1   import time
    2   from adafruit_circuitplayground.express import cpx
    3
    4   while True:
    5       x, y, z = cpx.acceleration
    6       print((x, y, z))
    7       time.sleep(1)
```

Adafruit CircuitPython REPL

```
Press any key to enter the REPL. Use CTRL-D to reload.
Adafruit CircuitPython 5.2.0 on 2020-04-09; Adafruit CircuitPlayground Express with samd21g18
>>> import time
>>> from adafruit_circuitplayground.express import cpx
>>> print("light:", cpz.light)
Traceback (most recent call last):
  File "<stdin>", line 1, in <module>
NameError: name 'cpz' is not defined
>>> print("light:", cpx.light)
light: 6
>>> print("light:", cpx.light)
light: 307
>>> cpx.pixels[0] = (255, 0, 0)
>>> cpx.pixels[1] = (0, 255, 0)
>>> cpx.pixels[2] = (0, 0, 255)
>>> cpx.pixels[3] = (255, 255, 255)
>>> cpx.pixels[4] = (100, 0, 100)
>>> cpx.pixels.brightness = 0.5
>
```

D

my torch (flashlight, for those of you across the pond) on the sensor to see how it would change.

The last six lines of code are me playing with NeoPixels — smart LEDs that let you change colors, adjust brightness, or even make animations. There are ten NeoPixels on the Circuit Playground Express, numbered 0 through 9. In the line **cpx.pixels[0] = (255, 0, 0)** I am setting NeoPixel 0 to red. Changing the number in the square brackets changes which NeoPixel you're addressing. Changing the values separated by commas in the brackets changes the color mix of red, green, and blue, from 0 through 255. The value **(255, 0, 0)** maxes out the red value and zeros out the green and blue, giving me a pure red color. The next four lines are me playing with colors until I find a shade I like (purple!). My final line sets the brightness of all the NeoPixels to **0.5**. The maximum brightness is 1, with 0 being off.

REPL makes it super fast and easy to tinker with settings and experiment with code. It's great for trying out settings, playing with sensors, and working through problems. Once you know your way around the REPL you'll never prototype another way!

E

F

Adafruit

Python Projects in the Wild

There are many talented people in our community using Python to make awesome things. A fun, nostalgic example is Scott Shawcroft's use of Circuit Python to hack old Game Boys (Figure **E**) using a game cartridge PCB with a SAMD51 chip (Figure **F**).

Another project using the SAMD51 is Bats 'n Gons from Sophi Kravitz and Ted Yapo, a super cool bat soundboard featuring emergent LED and

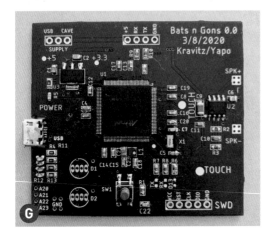

sound behavior: Each bat-shaped board can plug into the next to communicate over RX/TX, creating changes in light and sound output that change with each new board that is connected (Figure).

Into modular synths? You absolutely must put two projects from Thea Flowers at Winterbloom on your wish list. Sol (Figure **H**) is capable of translating MIDI to Eurorack-compatible frequencies and because it uses CircuitPython, it's easy for even beginner programmers to create and customize their own sounds. Big Honking Button (Figure **I**) is a very funny modular synth homage to everyone's favorite naughty goose (*honk honk!*). Both projects make use of Circuit Python's existing hardware and software to simplify their manufacture and code base: the Big Honking Button, for example, uses only 100 lines of code.

If you're up for a technical challenge you can check out the open source projects Migen and LiteX, which take Python's crossover into hardware to another level by exploring the ways in which the language can be used in FPGA and chip design.

And GreatFET (Figure **J**) is a fantastic project from Michael Ossman and the hardware hackers at Great Scott Gadgets. They take Python implementation in another direction by focusing on peripheral development rather than embedded development. Michael explains: "We believe that the best tool for rapid development and prototyping is not making firmware easier but making firmware unnecessary." With its Hi-Speed USB host and peripheral ports, and a Python API, the GreatFET One board can serve as a custom USB peripheral or a "man in the middle" making it a versatile interface for reverse engineering, hacking, and physical computing. ●

Microcontrollers, Meet
MicroBlocks

Physical computing software for education that runs live, on the microcontroller! Written by Kathy Giori

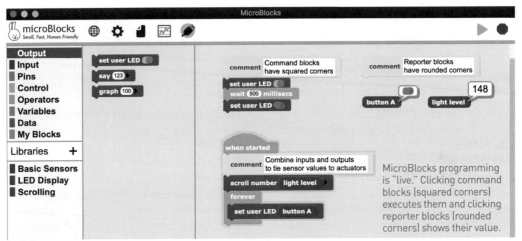

MicroBlocks programming is "live." Clicking command blocks (squared corners) executes them and clicking reporter blocks (rounded corners) shows their value.

MicroBlocks (microblocks.fun) is a coding environment for microcontrollers that's human friendly, but more than that, it's education friendly. Founder John Maloney worked on Scratch for a decade; his sidekick Bernat Romagosa founded Snap4Arduino, and the third musketeer, Jens Mönig, heads Snap! development, so MicroBlocks and Snap! can communicate using standard web services.

I got involved with MicroBlocks when I was developing the WebThings Gateway by Mozilla (see *Make:* Vol. 72, "Your Own Private Smart Home," makezine.com/go/private-smart-home).

Improve Learning by Coding Things "Live"

What's so cool about MicroBlocks? It's similar to Scratch and Snap!, where command blocks (rectangular) can be executed, and reporter blocks (rounded) can be evaluated, simply by clicking on them. MicroBlocks works the same way, except your commands control a microcontroller that can interact with the physical world. You can click a block to turn on LEDs, drive servos and motors, and play music. You can sense inputs from buttons, temperature sensors, accelerometers, and so much more. MicroBlocks scripting offers students what they expect — coding interaction with the physical world that's live.

Other blocks-based physical computing environments, such as Snap4Arduino and the Scratch micro:bit extension, support live programming by running your code on a PC and sending commands to a microcontroller. But the microcontroller does nothing when detached.

With MicroBlocks, your code runs directly on the microcontroller. As you work, your code is incrementally compiled, downloaded, and stored in persistent Flash memory so it still works even when the microcontroller is untethered and powered by a battery pack. This portability

Kathy Giori has developed open source hardware and software at Mozilla, Arduino, Qualcomm, and various startups. She holds EE degrees from Minnesota and Stanford, and authored "Your Own Private Smart Home" in *Make:* Volume 72.

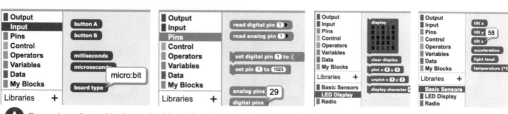

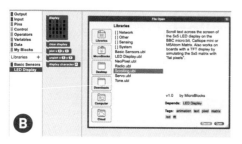

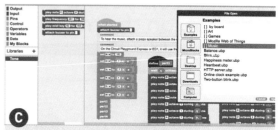

A Examples of core blocks and add-on libraries available in the blocks palette.

B

C

and autonomy allows MicroBlocks projects to be worn, carried in a pocket, or deployed to collect data outdoors.

When MicroBlocks detects which board is connected to your PC, it will automatically pull in libraries to support the features of that board. For the BBC micro:bit, two libraries are automatically added to the blocks palette: Basic Sensors (reporter blocks) and LED Display (command blocks). When you attach external components (e.g., NeoPixels, or a distance sensor) you can manually add libraries to support those components. Figure A shows two core categories (Input and Pins), and two from added libraries (Basic Sensors and LED Display).

These libraries make physical computing so much easier. Trying them out is just a few clicks away. Figure B shows the process of selecting and adding the Scrolling library.

For sensors and actuators external to the board, simply enter the pin number associated with the attached component. For example, `Attach buzzer to pin (14)` sets up your board to play musical tones or even songs, on a piezo speaker attached to pin 14. A favorite built-in example among students is the theme song for Harry Potter. Figure C shows how easy it is.

Different categories of blocks are color coded, making it easier to figure out what the blocks do. Core categories also include:

- **Control:** "when" hat blocks, conditionals and loops, time delays, broadcasts, comments
- **Operators:** math, expressions, logic

- **Variables:** to create local or global variables and set or change them
- **Data:** list and string functions
- **My Blocks:** to create your own command or reporting functions.

Outside these core functions, the rest of the blocks and libraries typically relate to physical computing. Actuators are controlled as Outputs. Sensors are monitored as Inputs. Communication can be done over both wired and wireless channels including I²C, SPI, peer-to-peer radio, HTTP, Wi-Fi, and infrared remote control. Figure D shows a snippet from a fun built-in example called RadioTexting.

One of the coolest aspects of MicroBlocks is the ability to plot data in real time! It lets you understand the output of a sensor so you can create algorithms. Figure E shows a plot of **motion** sensed from a micro:bit accelerometer while a person was walking. This live plotting made it possible to create a Step Counting library.

Turning Microcontrollers into IoT Devices

How do we teach the next generation to create smart, truly connected devices? IoT vendors like to tout "end-to-end" security, but that means data from your device go directly to the cloud, which disables interoperability between devices on your local network. And what if internet access is down? Or the cloud provider goes out of business? Who else can access your data?

Fortunately, MicroBlocks offers simple

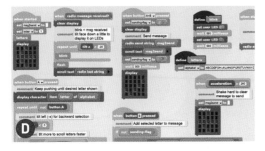

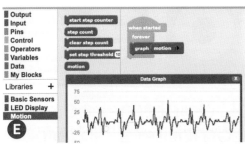

Kathy Giori

HTTP libraries and a Web Thing library that's compatible with Snap! and the Mozilla WebThings Gateway. Students can easily create their own IoT devices, using HTTP as a ubiquitous interoperability, control, and monitoring layer. These free, open source tools provide complete privacy and standard web security, with no dependency on big tech titans and no need to ask permission for cloud subscriptions, input a credit card, or give up privacy by sharing data.

To see some easy web devices made with the BBC micro:bit and WebThings Gateway, read more at makezine.com/go/microblocks.

Low-Cost Boards, High Educational Value

What hardware can you program? Anything that can run the MicroBlocks virtual machine (VM). Several pre-compiled VMs come built into MicroBlocks, and can be installed on your microcontroller board via a menu command. For less common boards, you can build and install the VM using PlatformIO (platformio.org). Many boards from the following processor families are supported (Figure F and Table, below):

- Microchip (Atmel) SAMD21 (Arm Cortex M0)
- Nordic nRF51, nRF52 (Arm Cortex M0, M4)
- Espressif ESP8266 and ESP32 (Tensilica Xtensa)

Other boards in these families are often easy to add. For example, dozens of ESP8266 boards can run the pre-compiled NodeMCU VM, although some use different pins for the user LED.

Get Started with MicroBlocks

1. Download MicroBlocks for Windows, Mac, Linux, or Chromebook at microblocks.fun/download and follow the setup instructions there.
2. You can also run MicroBlocks in a Chromium-based browser such as Chrome or Edge. To connect to your board via the WebSerial API, visit chrome://flags and enable the Experimental Web Platform Features flag. (Assuming WebSerial becomes a standard feature of Chrome, that step will no longer be necessary.) Then just browse to microblocks.fun/run/microblocks.html.
3. Once you've got MicroBlocks, see microblocks.fun/learn for next steps. For educators and makers alike, the site offers excellent Activity Cards and Guided Activities to get you started.

Ready to enjoy physical computing and have fun making IoT devices? MicroBlocks is the answer! ⊘

Boards supported by MicroBlocks	
Chipset core	**Example boards**
SAMD21	Adafruit — Circuit Playground Express (CPX), M0 models of Gemma, Itsy Bitsy, Trinket, and Metro Express
	Arduino — MKRZero, MKR1000, Zero/M0
	Microchip (Atmel) — SAMW25
nRF51, nRF52	BBC micro:bit, Calliope mini, Adafruit CPX Bluefruit
ESP32	Citilab ED1
	M5 — Stack, StickC, Atom
	IoT-Bus IO (discontinued)
ESP8266	NodeMCU
	Wemos D1 mini

Shut Your Pi-hole

Use a Raspberry Pi to block internet ads on your entire network — computers, smart TVs, phone apps, all of it **Written by the Pi-hole Dev Team**

 Pi-hole is open source software developed by a handful of volunteers with full-time jobs. The Pi-hole community shares skills, knowledge, and suggestions for new features at discourse.pi-hole.net, and we welcome everyone to contribute to issue reports and create pull requests at github.com/pi-hole.

What is Pi-hole?

Pi-hole is open source software that acts as a *DNS sinkhole* to protect your devices from unwanted content, without the need for any client-side software like apps or browser plug-ins. Put simply, Pi-hole acts as a mini DNS server that blocks your devices from looking up the internet addresses of advertising servers, so they can't serve the ads. This way, it blocks ads not only in browsers, but also mobile apps and smart TVs — and it makes your network faster! We call it a "black hole" for internet advertisements.

What is DNS?

At a very basic level *DNS (Domain Name System)* is like a phone book for domain names on the internet. It's what your computer uses to translate hostnames (e.g pi-hole.net) to IP addresses (in this case, 206.189.252.21). A request for an IP address is usually sent to an *upstream DNS server* (e.g., Google's 8.8.8.8).

How does Pi-hole block ads?

Pi-hole sits on your network in the middle of your computer and the upstream DNS server.

When Pi-hole receives a DNS request, it checks whether the domain exists on its blacklist, and if so, it returns the address 0.0.0.0, which is DNS-speak for "This number has not been recognized, please check the number and try again."

If the domain is not on Pi-hole's blacklist, then the request will be forwarded to the upstream DNS server, and the actual IP address will be returned to the client, your device.

What else can I block?

Pi-hole was designed to block known ad-serving domains, but really any domain can be added to its blacklist, and it will have the effect of the content not being served to the requesting client. So you can block known malware sites and other odious domains. We can think of a few.

SET UP YOUR PI-HOLE

Pi-hole can be installed on any hardware with a supported operating system (docs.pi-hole.net/main/prerequisites), although most commonly it is installed on a Raspberry Pi. Even a Pi Zero (Figure **Ⓐ**) will do!

TIME REQUIRED:
15–60 Minutes

DIFFICULTY:
Easy

COST:
$10–$50

MATERIALS
» **Raspberry Pi single-board computer** any model, from the Pi Zero ($5) to the new Pi 4 ($35), with power supply
» **microSD card, 4GB or larger** with adapter for your computer
» **Ethernet cable (optional)** if you want to plug your Pi-hole into your network. Of course you can also use Pi-hole wirelessly.

TOOLS
» **Computer with Balena Etcher software** free download from balena.io/etcher

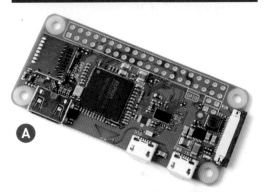

SPEEDRUN VERSION

1. Once you have your Pi up and running, simply open a terminal and run the following one-line command:

```
curl -sSL https://install.pi-hole.net |
bash
```

This will download and run the automated install script from github.com/pi-hole/pi-hole/blob/master/automated install/basic-install.sh. (If you're uncomfortable with directly piping shell scripts to Bash, there are other install methods available at docs.pi-hole.net/main/basic-install.)

2. From there, just follow the on-screen instructions (Figure ⓑ on the previous page) to get up and running. Now Pi-hole is installed and you are blocking ads!

NOOBS VERSION

1. Download the latest supported version of Raspberry Pi OS (aka Raspbian) (your choice of desktop or lite) at raspberrypi.org/downloads/raspberry-pi-os. As of this date, Buster is the newest supported version for Pi-hole.

2. Using Balena Etcher software on your computer, burn this downloaded image to your Raspberry Pi's microSD card, using a card adapter that fits your computer.

3. Set up the new Pi to enable ssh:
- Copy a blank text file named *ssh* to the boot directory of the microSD card mounted on your computer.
- For wireless access (optional), install a plain text file named *wpa_supplicant.conf* in the boot directory of the microSD card. Use the following text, but substitute your Wi-Fi network info:
```
country=US
ctrl_interface=DIR=/var/run/wpa_
supplicant GROUP=netdev
network={
    ssid="your-SSID-here"
    psk="your-login-password-here"
    key_mgmt=WPA-PSK
}
```

This will enable the Pi to join your home network.

After first boot, both of these text files are deleted and the configuration is saved in other locations on the card for subsequent boot-ups. If you're outside the U.S., see en.wikipedia.org/wiki/ISO_3166-1_alpha-2 for two-letter country codes.

4. Put the microSD card in the Pi, hook up the power to the correct micro-USB port (on a Zero, the left one is for data only), and plug it in. It should boot up and you should see the Pi on your network. If you don't, you can run Pi Finder (from Adafruit) and it will help you find the Pi's IP address on your network.

5. You can access the Linux terminal of the Pi via ssh from your client terminal:
ssh pi@<*IP-address-here*>
From there, you can continue setting up the Pi.

> **NOTE:** Putty is a Windows alternative for ssh.

6. Run **sudo raspi-config** and set up the localization: Wi-Fi in appropriate country, language and keyboard, time zone. The Pi will reboot after this. When it comes back up, shell in again and run **ip addr** to verify it has the correct assigned IP address.

7. Update the Pi OS to the latest version before you install Pi-hole. From the Linux terminal, run the following commands to update all the software packages to whatever's current:
sudo apt update
sudo apt upgrade

8. Now it's time to install the Pi-hole software:
**curl -sSL https://install.pi-hole.net |
bash**

> **NOTE:** Again, if you prefer not to pipe to Bash, see docs.pi-hole.net/main/basic-install for alternate install options.

The install command installs Pi-hole from the Git master branch. Follow the screen prompts, and when asked, enable the web interface. We recommend Cloudflare (1.1.1.1 and 1.0.0.1) as your upstream DNS servers to start. Turn on IPV6 if you use it on your network.

Pi-hole relies on third-party block lists. Select all the offered lists (these can be changed later) to add them to your Pi-hole's blacklist.

> **NOTE:** The Pi-hole documentation discusses the various upstream DNS servers so you can decide which is best for you (or choose any available server not on the list). docs.pi-hole.net/guides/upstream-dns-providers

9. After install is complete, verify that Pi-hole is running by going to the admin page: http://<*ip-address-here*>/admin or http://pi.hole/admin.

Ads-B-Gone

At this point you should have a fully functioning Pi-hole. In your router's DHCP configuration, point your DNS to the Pi-hole (and nothing else), and restart the router (Figure **C**). On your client devices (tablets, computers, phones), renew DHCP leases and/or clear DNS caches as necessary, and you're all set. You can also manually configure each device to use Pi-hole as its DNS server.

That's it. Browse, game, and binge ad-free! You can watch your Pi-hole work in real time using the web interface (Figure **D**). Go to the Dashboard tab to see total queries, blocked queries, total domains on your block lists, and graphs of the action over time.

Oh, the Things You Can Block

Pi-hole v5.0 was released in May 2020 with new features such as per-client blocking (choose which block lists to apply to which devices) and deep CNAME inspection to prevent domains being masked. And in v5.1 released in July, we added Dark Mode!

Pi-hole Away From Home

Pi-hole can also block ads to your mobile devices when you're away from home. By pairing your Pi-hole with a VPN to tunnel back to your network, you can have ad blocking on your cellular devices, helping with limited bandwidth data plans. Learn more at docs.pi-hole.net/guides/vpn/overview.

Watch Me Block It

A satisfying way to extend this project is to add a display right on the Pi. Adafruit built a Pi-hole with a tiny monochrome OLED or mini color TFT (learn.adafruit.com/pi-hole-ad-blocker-with-pi-zero-w) to display the number of ads blocked, DNS queries handled, and client devices benefiting from the blockage (Figures **E**). Later they upgraded the project (learn.adafruit.com/pi-hole-ad-pitft-tft-detection-display) with a bigger 3.5" color TFT and PADD client software (github.com/jpmck/PADD) that displays way more information (Figures **F**), including total domains blocked, total queries Pi-holed, and top offending ad domain. ●

Pi-hole, engage! Ads-b-gone.

C

D

E

F

Adafruit

NEW KID ON THE BLOCK:
Anne of All Trades started
her own YouTube channel
in 2018.

LIGHTS... CAMERA... MAKE!

These standout creators offer advice for putting your projects on YouTube Written by Mike Senese

There are over 300 diverse hours of footage uploaded to YouTube every single minute. A portion of this comes from makers around the world, some having refined their output into regular and frequent uploads that highlight their individual styles of builds and onscreen presentation. Done well, their videos inspire viewers to pick up their own tools, and also to subscribe to and share their creators' channels. The videos may also move viewers to grab a camera and start documenting their work as well.

If you're looking to become a regular content creator, there are many aspects to keep in mind. We've asked some of the top YouTube makers about their journeys, and for insights that might help you on yours.

GETTING STARTED
One of the hardest parts of any new venture is finding the courage to take the first step.

"Everyone has something good to share, and there is always an audience for it," says **Electroboom's Mehdi Sadaghdar**. "Always do what you enjoy, because that will guarantee you will do it better."

"When we were kids, you had to be a professional to get a book published, to be on TV, to get a record published. To be seen you had to be on a certain level," says **Bob Clagett** of **I Like to Make Stuff**. "And now, you can do all those things from your basement with nobody's permission."

Norm Chan (Figure **A**), who co-founded the video-heavy website **Tested** in 2010, offers three pieces of advice for those looking to begin. "One: Start. Two: As you make things, don't stop consuming, because you'll learn more from making while continuing to watch, and you'll learn new things you didn't realize were there. And Three: Write. If you write down what you say, you'll be able to communicate your idea in a clearer way than ever thought possible. Not even a script, just write your idea down."

"You're going to suck the first couple times, and that's completely OK. You're going to have to try out different routes, different personalities. Don't

Josh Nava, Mike Senese

"YOU'RE GOING
TO SUCK THE FIRST
COUPLE TIMES, AND
THAT'S COMPLETELY
OK." —*Ben Paik, Woby Design*

Michael Ryu @michaelryu, Mike Senese, Mehdi Sadaghdar, Jimmy DiResta

quit too early. Take your time and really do what makes you happy." —**Ben Paik, Woby Design** (Figure **B**)

GEAR
Fancy equipment can improve audio, open up new types of shots, and provide interesting effects. But if all you've got is a camera, you're still ready to go. (For more gear guidance, see page 48.)

"Over the past five years, I've just been using iMovie. I have an entry level DSLR on a tripod and I just move it around and capture what I'm doing," says **April Wilkerson** (Figure **C**). "It doesn't have to be extremely high-tech or complicated. I recently did a project and shot 90% of it on an iPhone 10. And people loved it. It's more about the storytelling and communication style."

"Learn to use your camera. Learning a few simple settings for white balance, light metering, and focal points can go a really long way in improving your content." —**Anne of all Trades**

"Doing YouTube is not about having good filmmaking tools, it's about the content you provide." —**Mehdi Sadaghdar** (Figure **D**)

"I try to keep it as simple as possible. It's really about the content." —**Ben Paik**

PRODUCTION
Experience will guide you as you figure out how to present yourself on camera, get shots that work, and avoid going crazy when you're piecing your clips together. Be ready to spend a lot of time on this.

"Treat the camera like a friend you are excited to share your passion with." —**Mehdi Sadaghdar**

"The biggest misconception people probably have is how much time it takes to do all the things." —**Jay, Wicked Makers**

"The fact that you're filming a project makes the project take five times as long. You're setting up the camera, then changing the angle, you're talking about the project, then re-recording what you said five times." —**Jaimie, Wicked Makers**

"Don't shoot a surveillance video when shooting your movie — get intimate with your point of view. Only shoot moments of transformation; don't shoot sanding for more than 15 seconds! Get in late, get out early ... in your edit."
—**Jimmy DiResta** (Figure **E**)

"I learned to shoot only what was necessary, and that made the editing easier later on."
—**David Picciuto, Make Something**

"It takes about at least a week, 6 hours a day, for me to edit these videos." —**Ben Paik**

"Typically when you see me build something, it's for the very first time." —**April Wilkerson**

"I do live music in the street for my vlogs," says **Jimmy DiResta** about finding royalty-free music. "And the other thing I do is just go to YouTube Studio. The file just downloads as an AIFF or MP3, and I drag it from the downloads folder right into my timeline."

STOKING CREATIVITY
The first videos you make may come easy, but you'll need to start planning your content to keep a flow.

"Let ideas roll around in your head for a few days before you sketch them," says **Jimmy DiResta**. "They need to incubate in your imagination before you commit them to paper. And keep a notebook on your workbench."

Homesteading-focused **Anne of All Trades** recommends developing a mission statement to keep focus: "Mine is, 'To share joy whenever possible. To learn to do or source as many things as I can myself or from within my local community. In the process, I want to encourage, inspire and empower my peers to do the same.' If a project or potential sponsorship or storytelling process doesn't communicate those goals clearly, it's not worth my time or my energy because it won't serve my business well long-term by distracting in any way from those goals."

"I'm planning or building my videos all the way through 2021." —**Mark Rober**

E

"I'm legitimately doing the same exact formula that I started with. I just wake up on Monday and think 'What do I need?' and then I go figure out how to build it." —**April Wilkerson**

CHANNEL GROWTH
Riches await those who hit a million subscribers. Or at the least, a framed golden YouTube badge. Most creators offer the same advice: Produce regular, consistent content, and buckle up for years of slow growth.

"The consistency of just sticking with it, it's a groundswell thing," says **Bob Clagett**. "Even though the algorithm is changing all the time, once you get traction, it gives you traction, which gives you traction."

"My idea was to just create really good content, regardless of what I am doing," says **Ben Paik**. "If my work is good then it is going to eventually catch on. At the same time I had a lot of help with different makers that shout me out, or do a little bit of collaboration. The maker community has been really generous to open their arms. They've been really open and really supportive."

"SUBSCRIBERS IS A VANITY NUMBER. IF YOU'RE ONLY GOING AFTER THE NUMBERS, YOU'RE GOING TO LET THE CONTENT LEAD YOU OFF THE PATH." —*April Wilkerson*

"There's no magic formula to be successful on YouTube. Make videos people want to share and growth will naturally happen over time." **—David Picciuto** (Figure **F**)

"A growth mindset on YouTube is, you should be making stuff not for your current subscribers, but for the people who haven't subscribed to you yet. You're trying to convince them to subscribe to your channel." **—Mark Rober** (Figure **G**)

GOING PRO

Sponsored videos are a common source of YouTuber income, as are affiliate links, AdSense, selling plans and merchandise, offering courses, hosting a Patreon, and more. On the other side, marketing, accounting, payroll, and management are all serious jobs, so be ready for that too.

"The sponsors are distinctly the best source of income for makers, I think everyone agrees on that," says **Jay** of **Wicked Makers** (Figure **H**). "But it's also a tremendous amount of effort to do it properly. You've got to build relationships, you've got to produce content that's going to be helpful for the sponsor. Or you can drop an affiliate link in the description of a video, and that isn't going to help you nearly as much as a sponsor relationship, but it's also no extra effort to do."

"When it comes to making videos, I focus on sponsorship because I can't really manufacture things out of skateboards and make money," says **Woby Design's Ben Paik**, of his skateboard-wood-focused channel.

F

G

H

I

Dan Struffolino, Mike Senese, Wicked Makers, Marnie Clagett

"The problem with just making videos is it's a one and done thing." says **April Wilkerson**, who has set up a full range of income sources. "You can do some affiliate links or a sponsored video but those are all short-term revenue streams. The website is the thing that is needed and crucial for having a long-term business. Making the videos is a good launching point, but it's very short minded and short lived."

"I just quit my job at Apple six months ago," says **Mark Rober**, one of the bigger YouTube successes of any genre. "It was four and a half years before I even realized you could make money off YouTube. Please don't start making videos because you want to become rich and famous. Make videos because you want to share your passion, and because you want to get better at making videos, and you want to get better at making things."

BURNOUT
The pressure to constantly be capturing and posting content can drain a creator. Here's some advice to keep that from hurting you.

Bob Clagett (Figure 1) started **I Like to Make Stuff** in 2012, which has now grown to a four-person team encompassing two video channels, classes, merchandise, podcasts, and more. It all began to take a toll last year. "We spent time talking about how we could lighten the load and avoid the bottleneck, because I am the bottleneck in all this stuff. We came up with ideas of different types of content, but instead of sticking them in between project videos in the same schedule, we've replaced one project a month with a non-project video."

"Foster good habits. Do some things just for fun, with no camera around. Get enough sleep. Take time off. Don't get burnt out."
—Anne of All Trades

"I hired a videographer and editor this year. That's two days every single week of my time that I'm no longer going to have to do." **—April Wilkerson** ⊘

CONTRIBUTORS:

ANNE OF ALL TRADES
youtube.com/anneofalltrades
Launched: 2018
Subscribers: 164,000

APRIL WILKERSON
youtube.com/AprilWilkersonDIY
Launched: 2013
Subscribers: 1.24 million

BEN PAIK — WOBY DESIGN
youtube.com/wobydesign
Launched: 2016
Subscribers: 126,000

BOB CLAGETT — I LIKE TO MAKE STUFF
youtube.com/iliketomakestuff
Launched: 2013
Subscribers: 2.98 million

DAVID PICCIUTO — MAKE SOMETHING
youtube.com/makesomething
Launched: 2013
Subscribers: 662,000

JAIMIE AND JAY — WICKED MAKERS
youtube.com/wickedmakers
Launched: 2018
Subscribers: 77,000

JIMMY DIRESTA
youtube.com/jimmydiresta
Launched: 2007
Subscribers: 1.79 million

MARK ROBER
youtube.com/markrober
Launched: 2011
Subscribers: 12.8 million

MEHDI SADAGHDAR — ELECTROBOOM
youtube.com/electroboom
Launched: 2012
Subscribers: 3.66 million

NORM CHAN — TESTED
youtube.com/tested
Launched: 2010
Subscribers: 5.08 million

TOOLS of the TRADE

Set up your YouTuber studio smartly with these tips

Written by Mike Senese

If you want to get going on YouTube, you're going to need some gear. You probably have what it takes to get started in your pocket right now, but as you grow into your channel, you might want to update some pieces to give you higher quality output and more control. Here is a pathway to help get you started, along with some tips and tricks for getting good results.

STARTER SETUP

Beginning with video? Just use your phone (Figure **A**). If it's anything from the past four years, it's going to do a great job (bonus: the built-in editing and uploading functions are maybe even better than a desktop). Don't let your gear hold you back — go to your shop and start filming.

NEXT: ADD SOUND

The second thing that a lot of creators focus on is audio. The classic adage is that people will overlook bad video quality, but bad audio will push them away.

The good news is you can get a $20 condenser lavalier microphone (Figure **B**) that will provide a sound improvement in many situations. The general style has an extra long cord (be careful to not trip over it) and uses a battery to provide power when used with those cameras that require it.

Putting the microphone close to yourself will help keep your voice clear and center, while minimizing further-away sounds. This setup works fine for a one-person presentation, but if you have two or more people, it gets trickier. We'll get to that in a bit.

The space you're in will also affect the audio. A big, spartan interior room will create a reverb-ed, echoey sound, which generally isn't too pleasing — it sounds amateurish and can make it hard to understand the person speaking. Closer microphone placement helps, but sometimes not entirely. To rectify it, you'll need to absorb the sound into something. Furniture can help — you may just want to go into another space with sofas and curtains to help soak up those echoes. Some people will get fancy and add foam sound panels on the walls and ceiling of their studio, but you can get the same effect by dragging a couple mattresses into your space and standing them up on their sides (Figure **C** , shown with pillow-booth,

MIKE SENESE is the executive editor of *Make:*, and has hosted television series on Discovery, Science Channel, and more. You can find his YouTube channel at youtube.com/mikesenese

Two quick beginner tips
• Keep yourself and your project in the center of the frame.
• Try to keep camera movement to a minimum. Even better, put the camera on something that doesn't move. A small cheap tripod is helpful (and doubles as a handle).

A

B

C

detailed next). Just try to keep them out of frame.

Another common trick for controlling your sound environment, especially for recording voiceovers, is to stand inside a closet while speaking. The fabric clothes and tiny space will give you a very dry sound.

In a pinch, you can also build a small one-sided vocal booth out of pillows or cushions on your desk (Figure **D**). Worst-case scenario? Hide under a heavy blanket while doing your lines.

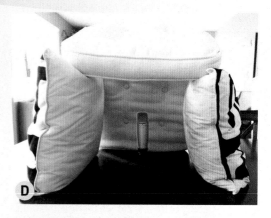

LET THERE BE LIGHT

If you're filming outside, lighting isn't a huge issue. Just be sure to not to stand between the camera and the sun, or you will be a silhouette (Figure **E**). Smartphones and GoPros are configured to handle everything else almost perfectly, and dedicated cameras have enough manual control to let you dial in the right shot too.

Indoor is trickier. If you've got big windows, you'll likely be fine with natural light during the day. But after the sun goes down, or in interior spaces without those big windows, the darker environment can be more problematic — forcing your camera to get into grainy ISO levels or just capping out so everything is a dark mush. This is where lighting kits come into play.

You can spend thousands on lighting gear, and can even dedicate your career to being a lighting professional (the same goes for any of the topics here). But there are a few tips for getting better, brighter shots without having to go that far.

First, turn on your lights. All of them. You may be surprised that even with what seems to be a bright lamp, your footage still seems dark. That's because our eyes are so dang good at compensating for dark environments, but technology is still working on it.

Next, try to avoid direct light — a flashlight or even headlights from a car will cast stark shadows. Instead, you'll want to diffuse your light over a large surface. Lampshades do this. Bouncing your light off a big white wall will do the same, although the bounce will eat up some of the brightness. If that's hard to do, just try to spread your lamps out around the subject so you really envelop it in light. Google "three-point lighting" for a more advanced approach to profiling a subject.

Mike Senese

WHAT ABOUT ACTION CAMERAS?

As mentioned previously, GoPro-type cameras can be an option, although we find them best for outdoor environments. Their ultra-wide-angle lenses also give an odd look to a person being recorded. They're not bad for a second camera when getting into a two- or three-camera setup, being a bit more rugged (especially when in their protective case). They can be fun for close-up shots of cutting and drilling, for instance, or strapped to the lumber you've got on your roof as you drive home from the store.

You may want to get a low-cost lighting kit. These run about $100 for a three-light set, and can be configured in a variety of ways. Or you can build your own — there are tons of step-by-step options online (see "DIY Gear" on page 53).

BACK TO THE CAMERA

If you're this far along, you may be getting ready to move past your phone camera. There are a few reasons to do this: more settings control, better resolution, adding new lenses, swappable storage and batteries. You might want to put the phone to use for other reasons (like recording audio, as we'll note in a moment). Or maybe you're just tired of getting phone calls in the middle of your best takes.

At this point, people tend to move up to DSLR and mirrorless options (Figure **F**). You'll have to spend a few hundred dollars for one of these at the entry level, and you might not even notice the difference most of the time. The convenience factor will also go down a bit — they're bulkier, the preview screens are not always visible from the front, and the footage can be harder to deal with too. They're definitely harder to use in an on-the-run, handheld selfie video clip moment.

But the flexibility can be worth it, especially when you get into manual exposure and focus situations, or upgrade to high-quality lenses that can give your scenes that tasteful blurred background look (called *bokeh*) that keeps the main subject clearly as the primary point of focus for the viewer.

This is also the time to get a new tripod (Figure **G**). Check Craigslist and other secondhand marketplaces for good deals.

ADVANCED AUDIO TIPS

Now you're moving into the deep stuff. Again, when it comes to video it's all about audio. Advanced audio often uses a dedicated recorder rather than the one in your camera — the quality ones will provide better sound. But at the DIY level, even just using your phone (which may now be freed up with a dedicated camera in use) and a free recorder app will give you more control options. Slip your phone upside-down in a shirt pocket to mimic a lav mic, or just use that $20 corded lav you got at the start, connected to your phone in your pocket. Have two people appearing on camera? Have them both do the same thing, and you've suddenly got two-channel audio with levels you can adjust in post.

Most decent editing software will let you bring in multiple audio channels, and many even have a system to sync everything up with your video — no more hand clapping needed. We've found that moving audio files from a phone to a computer can be frustrating, but you'll figure out a workflow with practice, and it will become straightforward.

ADD SOME MUSIC

Music can be a nice touch for a video when used aptly. If you're trying to make money with YouTube, stay away from commercial tunes, as they'll be recognized automatically and any earnings diverted to the rights holder. Instead, look for royalty-free tracks: Freemusicarchive. org has a number of Creative Commons-licensed tracks, and YouTube has free options as well. This is also an opportunity to reach out to your friends in bands and ask if they'd let you use something

(be cool and buy them some beers at the very least), or brush off your rusty skills and make your own tracks. Hey, you're a content creator now. Time to create that music content too.

CONCLUSION
Our advice is to keep it as simple as you can — the fancier you get, the harder and more time-consuming it will be to complete your video. Plus, equipment isn't cheap; there may not be a good return on investment from a dedicated piece of gear that does something only a little bit better than what you already have.

In the end, consistency is important, both with producing content regularly, and with giving that content a recognizable look and feel. But most crucial of all is the content itself — you can have the nicest video setup of everyone, but if your projects aren't interesting or useful, people won't come back. ◈

Mike Senese, Tyler Winegarner, Jeremy S Cook, Martin Taylor, Hep Svadja

DIY GEAR

Brownie Pan Light Panel
Build a few of these and you should be able to set up effective lighting on a budget. From *Make:* Volume 38.
makezine.com/projects/brownie-pan-led-light-panel

Noise Insulation Panels
Put these around your studio to suppress those pesky echoes.
makezine.com/projects/keep-it-down-build-noise-insulation-panels-for-your-shop-or-studio

Rolling Dolly
If your video style incorporates hero shots and lots of b-roll, a configurable roller dolly can add smooth pans and rotations to your options.
makezine.com/2010/09/10/pvc-skater-dolly

Motorized Slider
Great for putting motion into timelapse videos, but you can use these in a variety of other ways, especially when you're filming yourself alone.
makezine.com/projects/motion-control-camera-slider

iPad Teleprompter
For host-style camera shots, a teleprompter lets you use your script without looking away from the camera. This one puts your iPad to use for it.

Published in *Make:* Volume 60 (get a membership to access all the back issues at make.co).

YouTube
MAKER ALARM

Build a custom notification circuit that lights up
when your favorite maker uploads a new video

Written and photographed by Wesley Treat

WESLEY TREAT
is an author, maker, and
content creator, whose
projects he hopes will
inspire others to try new
things. You can find his
videos at youtube.com/
WesleyTreat.

When I had the pleasure of visiting the workshop of my buddy and YouTube personality Jimmy DiResta last fall, I came away with one of the numerous, plasma-cut logo stencils that tend to litter his workspace. I don't know exactly how it came into my possession, actually. It just sort of found its way in my bag. I really don't know how it got there.

At first, I was unsure what to do with it. But, when I got back to my own workshop, I devised a ridiculously obnoxious, Wi-Fi-enabled alarm that flashes DIRESTA in red and sounds an ear-splitting horn every time Jimmy uploads a new video to YouTube. Dubbed the DiRestAlarm, it's one of the dumbest things I've ever made. I couldn't be more proud.

I've developed a simplified version to share with you here. With a bit of existing Raspberry Pi knowledge and an internet-connected Pi at the ready, you'll be able to create your very own alarm to notify you when your favorite maker has published a new video.

BUILD YOUR YOUTUBER ALARM

1. WIRE UP THE CIRCUIT

The schematic for the simplified alarm is fairly straightforward (Figure **A**). The red LED, along with its resistor, is connected to the Raspberry Pi's GPIO via pin 18. The green LED, via pin 23. A momentary button, which will serve to reset the alarm once it goes off, is connected to pin 17.

The completed circuit, shown in Figures **B** and **C** on the following page, is really all you need to get started.

2. GIVE YOUR PI A GOOGLE ACCOUNT

To prepare your Raspberry Pi to receive YouTube notifications, start by going to Google and setting it up with its own account. The name of the account can be anything you want. This will create both a unique Gmail account and a login for YouTube.

3. TURN ON YOUTUBE NOTIFICATIONS

Log in to YouTube with the new Google account, then subscribe to the maker for whom you want to receive notifications. Make sure the bell icon next to the Subscribe button is set to All.

Next, go to Settings → Notifications → Email

TIME REQUIRED:
1–2 Hours

DIFFICULTY:
Intermediate

COST:
$50–$60

MATERIALS
» **Raspberry Pi single-board computer** with a Raspbian-flashed microSD card and internet access set up
» **LEDs: red (1) and green (1)**
» **Switch, momentary button**
» **Resistors, 330Ω (2)**
» **Solderless breadboard** for testing
» **Jumper wires**
» **GPIO breakout board (optional)** like the Adafruit Pi Cobbler — handy for prototyping, and for soldering the circuit permanently if you choose
» **Enclosure (optional)** of your choice

TOOLS
» **Computer keyboard, mouse, and monitor** for setting up the Pi; not necessary to run the alarm after that
» **Soldering iron and solder (optional)** to make the circuit permanent

Mark Adams Pictures

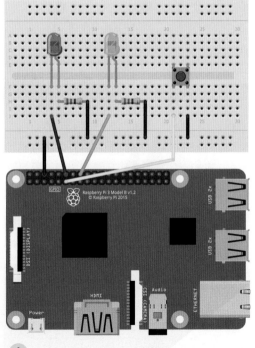

A

B

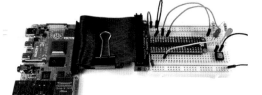

C

D

```
pi@raspberrypi:~ $ cd Downloads
pi@raspberrypi:~/Downloads $ python3 MakerAlarm.py
0 Unread Message(s) (2020 Mar 12 20:30:26)
Checking again in 09:58
```

E

```
pi@raspberrypi:~ $ cd Downloads
pi@raspberrypi:~/Downloads $ python3 MakerAlarm.py

1 Unread Message(s) (2020 Mar 12 20:34:06)
 - UID #4
   FROM: YouTube <noreply@youtube.com>
   SUBJECT: Wesley Treat just uploaded a video
New video!
Marking messages as read.

Checking again in 09:58
```

F

Notifications, and turn on "Send me emails about my YouTube activity." Also, make sure Subscriptions is selected under Your Preferences.

Now you should receive an email notification to the new Gmail account every time your favorite maker publishes a video.

4. INSTALL THE IMAPCLIENT LIBRARY

To more easily download and parse your email, we'll be using a library called IMAPClient, which you'll need to install on your Raspbian-running Raspberry Pi. You can install it simply by typing the following at a terminal command prompt:

pip install imapclient

5. PROGRAM YOUR PI

The real heart of the alarm is the Python script *MakerAlarm.py*, which you'll need to upload to your Raspberry Pi. The full code is available as a free download at makezine.com/go/ytmakeralarm.

Having a look at the script, a few variables may be of interest to you. **FROM** and **SUBJECT** are the attributes the script will search for when checking for new email. All other messages are ignored. **MAIL_CHECK_FREQ** defines how often to check for new mail. The default is **600** seconds, or every 10 minutes.

For testing purposes, you may change the **FROM** and **SUBJECT** variables if you would like to send your Pi a test email from your own personal email account. Just remember to change them back when you want to check for official YouTube notifications.

The only variables you need to customize are **USERNAME** and **PASSWORD**. **USERNAME** is simply the new email address you created above. **PASSWORD**, however, is not the password for your Google account, but a unique password you will need to create in the next step.

6. GIVE YOUR PI ACCESS TO GMAIL

Before your Pi can log in to check its Gmail account, it will need its very own password, or what Google calls an "app password." To set one up, log in to the new Google account, select Manage Your Google Account, and proceed to Security.

G

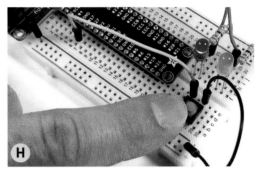

H

Under the section labeled Signing in to Google, select 2-Step Verification. You'll need to set this up before you can create app passwords. Simply follow the instructions and enter the code to verify that it's you.

Return to the Security page, where you should now be able to select App Passwords. Under the Select App option, choose Mail. Under the Select Device option, choose Other and enter a custom name, like "Maker Alarm." Finally, click Generate. Google will then create a 16-character password that you can use exclusively for the alarm. Copy this password before proceeding, because Google will not display it again.

Lastly, paste this password into the variable **PASSWORD** in *MakerAlarm.py* and save your changes. That's it.

MONITOR THAT MAKER!

To use your YouTube Maker Alarm, make sure your Pi is connected to the internet, then open a terminal window, navigate to the directory where you uploaded the script, and type the following:

`python3 MakerAlarm.py`

When the program starts, two things will happen: the red LED will turn on and the Pi will check its Gmail account. If it doesn't find an unread notification email from YouTube, the red LED will remain lit (Figure **D**) and the script will display a timer in the terminal window, counting down till it tries again (Figure **E**).

If it does find a notification email (Figure **F**), the green LED will flash several times, then remain lit (Figure **G**) as a reminder to let you know your favorite maker has uploaded a new video! Once you get a chance to watch it, simply

press the momentary button to reset the alarm back to red (Figure **H**).

Until then, the script will continue to check for new email in the background, and the alarm will flash again if a new, unread notification email is found.

AMP IT UP

Now that you've got the basic alarm working, you'll probably want to customize it. Create a nice enclosure by 3D printing a fancy case, or fashion one out of wood. You can amplify your alarm's effects by adding more or larger lights, or by incorporating audible alerts with a piezo buzzer.

Of course, if you want something that will really grab your attention, you can do like I did and hook it up to a startlingly loud motorcycle horn. (This involves a lot of power considerations, like a large 12V supply, a relay module to trigger the horn, and a step-down circuit to power both the Pi and the relay module.) But I don't recommend it if you're keeping the alarm in your workshop where you'll be operating any dangerous power tools! ◓

See the original DiRestAlarm build at youtu.be/3QGCLI-UFBA, and learn more about how IMAPClient works at imapclient.readthedocs. io/en/2.1.0

CAMERA SHY

The host of Ugly Duckling House
shares her contemporaries'
reasons for avoiding YouTube

Written and photographed by Sarah Fogle

YouTube simply wasn't on my radar when I started blogging 10 years ago. All the people I liked and followed were bloggers. I didn't watch YouTube videos to understand how to do something (my dad did, though). Instead, I scoured forums for people in trades and subscribed to bloggers recapping their projects. I *preferred* tutorials in their written form to digest at my leisure and repeatedly reference. As a result I began my blog by writing to someone just like me, and tried to give them what I came to blogs for (often reading from work, where I didn't want to get caught watching a video!).

"I'll spend extra time googling and looking around for a written- and picture-documented DIY and despair when I can only find videos."
—Olivia, @diyhorseownership

YOU GRAVITATE TOWARD SKILL SETS YOU ALREADY HAVE

I came from a customer service background in software where I did a lot of technical documentation. I simply "got" blogging more than I did on-camera work. As my blog became more popular, I went to conferences to network. The classes were focused on traffic, Pinterest, Instagram, and more, but never seemed to focus on YouTube.

PINTEREST HEAVILY INFLUENCED THE BLOGGER'S FOCUS

I hopped on Pinterest very early and it was a huge traffic driver to my blogs. It was a platform full of my peers (other bloggers I knew and URLs I recognized), so it was a natural link to add my content and encourage my followers to join. That symbiotic relationship was great for growth and encouraged my focus between these platforms. YouTube seemed like an outlier that had more obstacles to sharing as part of a social media strategy, especially with formatting (vertical for Pinterest, horizontal for YouTube).

IT'S A COMFORT ZONE THING

Many bloggers are introverts — comfortable sharing their passions and projects, but not as comfortable with a camera or speaking in public. I can write 1,000 words without batting an eye, but

SARAH FOGLE is a self-professed "power tool addict" who shares do-it-yourself tips, tutorials, and renovation realities at UglyDucklingHouse.com.

speaking into a camera or in front of a crowded room leaves me tongue-tied. That also means no need for makeup or changing clothes. I've heard some describe themselves as "lazy" when admitting this ... but to me, it sounds more like intimidation.

"I want my videos to all have the same aesthetic and intro, etc. and haven't figured all that out yet." —Charlee, @buildandcreatehome

And on YouTube, unlike a blog, *you can't go back and edit!*

For many, Instagram is a way to bridge the gap and build up confidence on camera. I tried videos out on Instagram first and saw it as a new way to demonstrate techniques and teach, to test the reception of how my followers would react to adding video, etc. Still, it requires more effort:

"YouTube requires a different video format than Instagram and IGTV, so I see it as more work getting content made just for YouTube." —Monica @house.of.esperanza

SPONSORS MAY OR MAY NOT WANT IT

For the majority of my blog's growth, sponsors and sidebar ad networks were all focused on blogs and my social media shares ... not videos. They never asked, and I never offered. No one seemed to talk about it until the last few years when the lines began blurring more between bloggers, IGers, and YouTubers. Still, I hesitated because ...

SOMETHING'S GOTTA GIVE

Once I had a successful blog, I had a full-time business. I had a pace for new content that readers could reasonably expect with small hints shared on Instagram to fill in content gaps. Adding YouTube to my workload would mean, somehow, adding content without compromising what I'd already built or alienating my existing followers. It seemed like doubling or even tripling the workload — and I was right!

"My main reasons are: 1. lack of time and 2. the lighting in my garage is horrible and using the natural light means additional editing as the

light source (the sun) is constantly moving."
—Char, @woodenmaven

"Time to learn a new platform, time to learn the editing tools, and then time to do the editing itself." —Chris, @ironhorse_woodcraft

"Lack of a decent PC to put together a proper video/tutorial." —@calibercreativeworks

"The thought of adding one more thing to my plate with work/life balance seemed really overwhelming to me. I didn't want to do it until I felt ready to put all of my energy behind it and do it right. I'm still figuring out how to balance it all, but I'm glad I finally got started."
—Erin, @erinspainblog

Once I launched on YouTube, the video skills needed were added to my pile of to-dos for each project. Costs increased. I still needed to stop and photograph certain steps. My total work time spent per project increased considerably, which meant spreading out my content calendar, and fewer projects could be completed per month.

DOES YOUR AUDIENCE EVEN WANT IT?
I also wondered, would my audience even respond? Or would this be a new audience — entirely separate from those I've reached before? Would quality suffer? Would I have to hire outside help to shed some of the extra work, and would that cost be balanced out by the views?

Now that I've had some success on YouTube, I know it does add value to my audience and makes more income, but it's still a *very* hard balance. Half of the videos I create for my blog still don't make it to YouTube because of the extra work required for intros/outros/voiceovers!

THEN THERE'S THE COMMENTS …
YouTube comment culture has a reputation that precedes the platform. A blog friend once called it "the outhouse of the internet," and I wondered whether it would be worth trying to grow a thick skin just to share the things I was passionate about.

It's not that I haven't received negative comments on my blog or Instagram before, and I still get plenty of positive comments now that I'm on YouTube. But the bad ones can be particularly cringey — they can feel more like attacks and sting a little bit more. Comments about my looks or body make me uncomfortable regardless of platform, and I get them more on YouTube. I found myself creating a blocked words list for the first time. Some content creators just don't have the mental space or energy for that, and I can't blame them.

"Honestly, I've seen a lot of my blogging friends deal with horrible comments and have no desire to bring that into my life. So it just never truly appealed to me to add a lot to YouTube."
—Sheri, @hazelandgolddesigns

With all of that said, now that I'm on YouTube I'm meeting an entirely new audience than I knew before, so I consider it value-add overall. It's been worth overcoming my assumptions, the workload, and all the times I've talked myself out of it.

I've met a ton of YouTube makers as well, which made me feel part of a new community where I can grow new skills (I'd never tried welding or blacksmithing before!). I've found a few new heroes; some have even become close friends.

Looking back, it flows along similarly to my experience with platforms before it — there once wasn't Pinterest, and I adapted; there once wasn't Instagram, and I adapted to that too; YouTube is my newest challenge, and isn't so scary, after all! ⊘

DIY Mobile Handwashing Station

Written by Sam Reardon with Kris Kepler, Colton Coty, Adam Vera, and Joey Freid

TIME REQUIRED:
1–2 Hours

DIFFICULTY:
Easy to Intermediate

COST
$300–$400

SAM REARDON is an East Bay Area native who works for LavaMae[X], a nonprofit dedicated to bringing handwashing to every community that needs it. A former preschool teacher, she has always wanted to teach, guide, and connect with others; LavaMae[X] has also allowed her creativity for projects to flourish.

Bring the clean — with this 32-gallon rig made from common hardware

In times like these you learn to ... lather up.
Everybody needs hand hygiene, everywhere. Most of us take it for granted but for some, it's just not readily available.

LavaMae[X] is a nonprofit accelerator changing the way the world sees and serves our unhoused neighbors. Our goal is to teach the world how to take critical services to the street by providing free toolkits, mentorship, and training for communities that want to launch LavaMae[X] designed programs such as shower trailers and DIY handwashing stations. To date, we have inspired or directly advised 189 mobile hygiene programs around the world.

Inspired by Love Beyond Walls, and a grassroots campaign between USC Annenberg and Los Angeles Community Action Network (LACAN), we created this mobile handwashing station, designed for longevity, durability, and high capacity while creating opportunities for community connection and ownership of the units. Our DIY 32-gallon trash bin design is simple enough for anyone to assemble using tools and parts easily found at a hardware store, and it's low-cost compared to commercial handwashing rentals. It can accommodate up to 500 handwashes at a time, and can be built, deployed, and maintained by the community.

Here's how to build one for your community.

BUILD YOUR MOBILE HANDWASHING STATION

This is how we do it. Be sure to watch the DIY video tutorial at youtu.be/zSe27aHez8c and visit our website lavamaex.org/handwashing-for-all before you build, for updated parts lists and instructions.

1. BUILD THE FRESH WATER BIN

1a. Decide where you would like the plumbing to come out of the lid of your bin. In our bin, we chose the center. Once you've decided the location, use a

Adam Vera, Sam Reardon, Javier de León, Colton Coty

MATERIALS

- » **32-gallon trash cans with lids (2)** Rubbermaid Brute #203118
- » **Trash can dollies (2)** Rubbermaid Brute #FG264020BLA
- » **Dishpan, 12qt, 15¾"×12½"×6"** Sterilite #06588012
- » **Vinyl tubing, ½" ID (10' length)** such as Everbilt #204698. It's more durable than garden hose.
- » **Galley foot pump, ½" fittings** such as Whale Marine #GP4618
- » **Hose clamps, stainless steel, ½"–1¼" (4)**
- » **PVC pipe, ¾" dia. (10' length)**
- » **Barrel bulkhead union washer fittings, ½" (3)** such as Everbilt #800469
- » **PVC reducer adapters, ½" male pipe thread (MPT) × ¾" slip (6)** Charlotte Pipe #PVC021100500HD
- » **Brass reducer adapters, ½" ID barb × ¾" MPT (2)** Apollo #APXMA3412
- » **PVC in-line check valve, ¾" slip**
- » **Plumbers tape, PTFE (Teflon)**
- » **PVC elbows, 90°, ¾": slip × slip (4) and slip × female pipe thread (FPT) (2)** such as Charlotte Pipe #PVC023000800HD and Dura Corp #C407-007
- » **ABS flood drain, snap in, 2"**
- » **ABS elbow, 90°, 2"**
- » **Electrical junction boxes, 4"×4"×4" (2)** Home Depot #R5133709. Highly recommended to hold your hand soap — it's more sanitary. These boxes hold up to 32oz. There are many alternatives, just make sure to use a closed container that holds a good quantity of soap and can be mounted onto the lid with bolts and nuts.
- » **Bolts, ¼-20 × 1", with nuts (4)** for mounting electrical boxes
- » **Clear PVC cement**
- » **Black ABS cement**
- » **Waterproof sealant** such as Loctite Clear Silicone
- » **Clear construction adhesive** such as Loctite Power Grab
- » **Hand soap, 16oz bottles (2)** We use The Right to Shower liquid soap, therighttoshower.com.
- » **Fresh Sink chlorine tablets**
- » **Paper towel holder** such as Kamenstein #4554ASB
- » **Hand sanitizer, 8oz bottles (4)**

TOOLS

- » **Ratcheting PVC cutter, 1¼"**
- » **Hole saw arbor, ¼"**
- » **Hole saws, ½" and ¾"**
- » **Drill/driver with drill bits**
- » **High-speed rotary tool with plastic cutting wheels (optional)** such as a Dremel. You can also use a hand saw for this.
- » **Hand saw**
- » **Screwdrivers, Phillips and slotted**
- » **Tape measure**
- » **Socket wrench with ¼" socket**
- » **Adjustable wrench** aka Crescent wrench
- » **Caulking gun**

½" drill bit or hole saw to cut the hole.

1b. Separate the two pieces of a ½" bulkhead union washer fitting (Figure Ⓐ), leaving one washer per piece. Push the piece with the threading through the hole you have cut and screw the other side on, on the underside of the lid.

1c. Wrap a few layers of PTFE tape (Figure Ⓑ) around the threaded areas of two ½"×¾" PVC male reducer adapters (Figure Ⓒ).

1d. Screw the PVC reducers into the top and bottom sides of the bulkhead fitting.

1e. Cut a 3" length of ¾" PVC pipe (Figure Ⓓ) and slide it into the PVC reducer that's on top of the lid.

1f. Slide a ¾" PVC elbow (slip × FPT) (Figure Ⓔ) onto the 3" pipe.

1g. Wrap a few layers of PTFE tape around the threaded areas of a ½" brass barb × ¾" MPT reducing adapter (Figure Ⓕ) and screw it into the FPT side of the elbow. Place the lid to the side.

1h. Measure the distance from the top of the bin to ½" above the bottom of the bin. Cut a ¾" PVC pipe to this length. Cut the bottom end of this intake pipe at an angle. You want it to stop ½" before hitting the bottom of the bin, to prevent sediment from being picked up by the pipe when pulling fresh water.

1i. Slide the non-angled end of the PVC pipe into the PVC reducer on the underside of the lid, then place the lid back on the bin.

TIP: To make sure the ¾" PVC pipe isn't hitting the bottom of the bin, press on the lid and try to feel if the pipe is hitting bottom. You can also try lifting the lid on one side and peeking in. If it's too long, simply cut it shorter with the PVC cutter.

2. DISHPAN CUTOUT AND FAUCET

2a. We recommend placing the 12qt dishpan (Figure Ⓖ) centered on the lid, but slightly forward, closer to where the guest will be washing their hands. This gives ample room for the dishpan and faucet. With a marker, trace the circumference of the sink basin on the top of the lid; you will be cutting along this line. Place the dishpan aside.

2b. Use a handsaw or a Dremel equipped with a plastic cutting wheel to cut along the traced line on the lid. You may need to trim the hole slightly a few times in order to get it just right. The dishpan will not be glued in, so the goal is to get it to sit comfortably and securely in the lid while allowing easy removal for access to eliminate the greywater that will fill this bin.

2c. Place the faucet to the rear of the dishpan. There should be just enough room; approximately ½" between the dishpan and the faucet (another ½" bulkhead fitting). Use the drill with a ½" hole saw to cut the hole for your faucet (Figure Ⓗ).

2d. Install the bulkhead fitting into the ½" hole you drilled.

2e. Wrap a few layers of PTFE tape around all threaded parts of two more ½"×¾" PVC male reducers, then screw one into the top and one into the bottom of the bulkhead fitting.

2f. Cut a length of ¾" PVC pipe at the desired height for your faucet.

2g. Slide this pipe into the PVC reducer on top of the lid.

2h. Slide a ¾" 90° PVC elbow (slip × slip) (Figure Ⓘ) onto the pipe.

2i. From the curved portion of elbow, measure

Adam Vera, Sam Reardon, Javier de León, Colton Coty

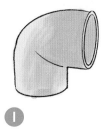

the horizontal distance to the drain, then use that measurement to cut a short length of PVC pipe.

2j. Slide this pipe into the elbow horizontally.

2k. Slide one last 90° PVC elbow (slip × slip) onto the open end of the horizontal pipe to complete the faucet.

3. INSTALL THE DRAIN

The drain we used (Figure **J**) separates into upper and lower pieces so that the dishpan "sits" between the two.

3a. Find the upper drain piece that will go through the sink basin, place that portion in the center of your dishpan, and trace its circumference onto the bottom of the dishpan.

3b. Use the handsaw or Dremel to cut along the traced line.

3c. Insert the upper part of the drain through the hole (Figure **K**), then screw on the lower part on the underside of the dishpan (Figure **L**).

3d. Slide the 2" ABS 90° elbow (Figure **M**) into the opening at the bottom of the drain, underneath the sink basin.

4. INSTALL THE PLUMBING

In our model, the plumbing exits at the front of the greywater bin. It's very important that the ¾" in-line check valve sits in a horizontal position. Your PVC piping also needs to go low enough in the bin to avoid touching the sink basin, drain, and elbow. Assuming your model will mirror ours, install plumbing as follows.

4a. Measure from the top of the sink basin to the lowest part of the 2" ABS 90° elbow.

4b. Add 1" to 2" to this measurement, and cut a piece of ¾" PVC pipe. Insert that piece into the PVC reducer that's screwed into the bulkhead on the underside of the lid.

4c. Slide a ¾" 90° PVC elbow (slip × slip) onto the end of the PVC pipe. The open side of this elbow should be facing the direction in which the plumbing will exit the unit.

4d. On the front of your bin (Figure **N**) where the plumbing will exit, use your measurement from step 4b to measure from the top of the bin, down toward the ground. Place a mark with a marker.

4e. Tip the bin on its side and drill a ½" hole at your mark.

> **TIP:** To prevent the bin from moving when drilling, position the bin so that the opening is facing you, and use one foot to step on the inside of the bin.

4f. Install the remaining ½" bulkhead into the hole. Wrap a few layers of PTFE tape around the threads of the remaining two PVC male reducers and screw one into either side of the bulkhead.

4g. Cut a 3" length of PVC pipe and slide it into PVC reducer on the inside of the bin.

4h. Slide the ¾" in-line check valve (Figure **O**) onto this pipe.

> **IMPORTANT:** There's an arrow on the in-line check valve; make sure this arrow is facing the faucet. The purpose of the valve is to prevent backflow of water inside the unit. If you install it backward it will prevent water from entering the bin properly and coming out of the faucet.

4i. Measure from the center of the check valve to the curved portion of the elbow from step 4c.

4j. Using that measurement, cut another ¾" PVC pipe.

4k. Slide one end of this pipe into the check valve, and the other end into the elbow from step 4c. This completes the plumbing inside the greywater bin.

NOTE: If the ¾" PVC pipe from this step is too long, trim it down a bit to ensure a proper, comfortable fit between all connections and pieces (Figure P).

4l. Cut a 3" pipe and slide it into the PVC reducer that's screwed into the bulkhead outside the bin.
4m. Slide a ¾" 90° PVC elbow (slip × FPT) onto the 3" pipe.
4n. Wrap a few layers of PTFE tape around the MPT threads of the remaining brass barb reducer and screw it into the FPT end of the elbow.

5. MAKE THE SOAP DISPENSERS

5a. On the lid of your freshwater bin, decide where you'd like to place the two electrical junction boxes (Figure Q). Make sure they're level, and accessible to the user washing their hands. We recommend placing them on the side nearest to your handwashing / greywater bin.
5b. With the boxes in place, use a marker to mark through the two mounting holes for each, then place the boxes to the side.
5c. Drill four ¼" holes through the lid at your marks. Make sure the ¼-20 bolts fit through the holes.
5d. Put the boxes back on the lid in the correct position, making sure the mounting tab holes line up with the holes in the lid.
5e. Push the four ¼-20 bolts (Figure R) through the attachment tab holes and the holes in the bin lid.
5f. Take the lid off of the bin and turn it sideways, so you can screw the nuts (Figure S) onto the bolts. Tighten as much as you can by hand.
5g. Use a wrench to hold the bolt head (above the lid) and a socket wrench with ¼" socket to tighten the nut (underneath) as much as possible.

NOTE: We recommend nuts and bolts instead of glue for durability. You can add washers inside to make it even stronger.

5h. Stand your soap dispensing pump (Figure T) inside the empty electrical junction box to determine how much, if any, of the straw on the pump needs to be cut. We removed about 2" of the straw from The Right to Shower dispensing pump so that the straw stopped ¼" above the bottom.

5i. Drill a ¼" hole through the center of the lid of the junction box lid to allow placement of the soap pump, then glue the pump in place using a caulking gun and clear silicone adhesive.

5j. Once the adhesive fully dries (check instructions), fill the junction box with 32oz of soap. Then, using a Phillips screwdriver or bit, tighten all four screws to attach the lid onto the box (Figure U).

6. CONNECT THE BINS VIA GALLEY PUMP

6a. With the PVC cutting tool, cut a length of vinyl tubing (Figure V) to go from the brass barb on the front of the greywater bin to the galley pump (Figure W) on the ground.

6b. Slide two loose hose clamps (Figure X) onto the tubing and let them stay loose for now.

6c. Attach one end of the tubing to the brass barb on the greywater bin, and the other end to the galley pump *outlet* connection.

6d. Slide two loose hose clamps onto the remaining tubing. Slide this tubing onto the *inlet* portion of the galley pump, and onto the brass barb on top of the freshwater bin.

6e. With a slotted screwdriver or bit, tighten one hose clamp over each barb and pump connection.

7. TEST AND GLUE

7a. Test your handwashing station by putting water in the freshwater bin and washing your hands. Make sure all parts and plumbing are working properly, and check for leaks (Figures Y and Z).

7b. After everything operates perfectly, apply PVC cement glue to all slip joints (Figure Aa); for example, where a ¾" PVC pipe slides into a PVC reducer or into the check valve.

7c. Apply ABS cement to the portion of the 2" ABS

Amber Wise, LavaMae^X, creating a handwashing station in Los Angeles.

elbow that slides into the sink drain.

7d. Apply clear silicone caulking to *both sides* of the washers/gaskets on the bulkhead fitting at the front of the greywater bin, for a leakproof seal.

7e. Put each bin on its dolly (Figure Bb) and you're done.

A LATHER OF LOVE

Communities around the world need broader access to handwashing and mobile sinks. Encampments, reservations without plumbing, rural outdoor living areas, shelters, churches, and city parks are just a few examples. Our goal is to meet those needs, or to be the hand that helps someone else do so. We're all in this together.

Please share your story with us! We'd love to hear your experience of implementing this project in your community. Questions, concerns, and suggestions are always welcome at lavamaex.org/handwashing-for-all. ⊘

Storm Globe

Forget boring snow globes — build a swirling tempest for your desk using magnetic stirring and LED lightning!

Written and photographed by John Thurmond

Are you the kind of person who likes a good storm? Staying inside and warm while lightning flashes and the wind lashes the rain around is something I have always enjoyed. What if you could capture that feeling in a sphere that you could keep on your desk?

SEEING TURBULENCE

Rheoscopic fluids allow the currents and turbulence in liquids to be seen. They're typically made with mica, a mineral that forms small, shiny flat plates that easily move within a fluid. The reflection of light off the mica turns the turbulence into a mesmerizing display.

You can find large spinnable round vessels filled with rheoscopic fluids in science museums (such as the Glasgow Science Centre) to demonstrate atmospheric flows over the surface of a planet, or as art displays (most famously, the Kalliroscopes of Paul Matisse), or even in your living room (the Rheoscopic Disc Coffee Table by Ben Krasnow, *Make:* Volume 47, makezine.com/projects/rheoscopic-coffee-table).

Making a rheoscopic fluid is simple, and it's easy and inexpensive to buy the mica flakes, because they're used in manufacturing soap, bath bombs, and makeup. I thought it might be interesting to try to shine a light from behind a rheoscopic fluid display, rather than just relying on reflected light to flash off the mica. A quick test showed this to be an interesting effect (see Video Resources).

The turbulence reminded me of a storm, so I thought it would be fun to bottle some lightning as well!

BOTTLE YOUR OWN STORM
1. MAKE A MAGNETIC STIRRER

Take two small magnets and stick them together. Since like poles repel, and opposites attract, this will let you identify the opposite poles of the two magnets. Use a marker to mark the opposite sides of each magnet (Figure A).

Then glue the magnets down to the edges of the middle of the fan hub (Figure B on the following page). Be sure to do this on the side that spins (the one without the label).

TIME REQUIRED:
A Weekend

DIFFICULTY:
Intermediate

COST:
$50–$60

MATERIALS
» **Circuit Playground Bluefruit microcontroller board** Adafruit #4333, adafruit.com. Other Circuit Playground boards should work with minor code modifications.
» **Micro-USB cable**
» **Fan, 40mm×40mm×10mm**
» **DIY Snow Globe Kit** Adafruit's 108mm version, #3722, fits the Circuit Playground perfectly, and was also included in AdaBox014.
» **Colored mica** Typically sold for makeup, soap-making, or bath bombs — a very little will go a long way!
» **Rare-earth magnets, 3mm×1mm round (2)** Any small strong magnets would likely work fine.
» **Magnetic stir bar, 15mm round** aka stirring flea. Other shapes and sizes should work as well or better.
» **Screws and spacers, M2.5 or M3, non-ferrous (brass or plastic)**
» **Cyanoacrylate glue**

TOOLS
» **Computer**
» **Wire cutters and strippers**
» **Marker**
» **Glue**
» **Soldering iron (optional)**

JOHN THURMOND
is a professional geologist and an amateur at many things. He's passionate about designing ridiculous gadgets and learning through failure. He'd love to hear your stories about either one of these @grajohnt!

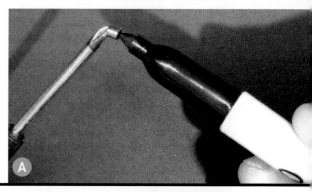

A

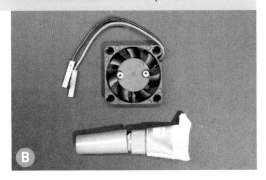

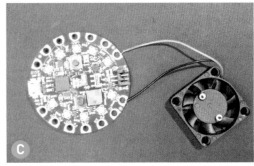

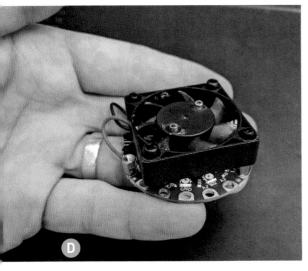

2. WIRE UP THE BLUEFRUIT

We're going to just use the USB voltage (5V) to power our fan. There are better ways to do this, but let's keep it simple! Cut off any connector on the fan wires, and strip the wires back a bit. Then attach the black wire to GND, and the red wire to VOUT (which is 5V, if you're powering the board from USB). You can solder these if you like, or just make sure the wire is wrapped around the edge of the hole tightly to make a good electrical connection (Figure C).

3. ASSEMBLE THE STIR PLATE

Attach the fan to the Bluefruit using non-ferrous M2.5 or M3 screws and standoffs. On a 40mm fan, the diagonal holes line up perfectly with two of the holes on the Bluefruit, which is convenient. You may need to vary the length of standoffs or screws to get an offset between the fan and the base of the snow globe, but the height of the screw heads should be enough space to allow the fan and magnets to spin (Figure D).

4. PREPARE THE SNOW GLOBE

Fill the snow globe with water, and add a magnetic stir bar — the little white lozenge shown in Figure E. A small bowl can help you keep the snow globe in place, or you can have a friend hold it for you.

It is very easy to add too much mica to your snow globe, which will make a nice swirly effect, but will be completely opaque to light. If you wet the end of a toothpick, dip it in the mica, then into the snow globe, and repeat that process a few times, you'll have the right amount of mica (Figure F). Some experimentation may be required, depending on the mica you use.

Place the plastic stopper on the snow globe. If you overfill it a bit, and tilt it while you slowly squeeze the lid on, you can get all the air out to avoid any bubbles.

It's likely there won't be enough room to attach the screw-on base; I didn't use it at all. If you're worried about leakage, you can glue the plastic stopper on.

5. PREPARE A BASE (OPTIONAL)

Your Storm Globe will work fine without a base. But I decided to make a Mars-themed version, so I designed a base from the 3D model of the Block Island rock mapped by the Opportunity rover on Mars, and 3D printed it. The STL file is available at github.com/grajohnt/StormGlass.

6. CODE THE BLUEFRUIT

There's an excellent Adafruit Learning Guide that explains everything about getting a Bluetooth Snow Globe up and running. Read it at learn. adafruit.com/snow-globe-bluefruit-cpb. Then download the Storm Glass code from github. com/grajohnt/StormGlass; it's a lightly modified version of their demo code.

The stormy section of the code lights up the LEDs on the Bluefruit in such a way that it looks like a **lightning** flash. This has some **random-** ness included in it, so no two lightning flashes are the same (Figure).

7. START YOUR STORM!

Magnetic stir plates typically include a speed control, which allows you to start the stirring process gradually. This one doesn't, so you must be careful to get your stir bar stirring correctly.

If you look underneath the snow globe, you should be able to see your stir bar. Tilt the globe around until the stir bar is more or less in the middle, and then place it on top of the assembled base. Then plug the Bluefruit into a USB port to start the stirring effect. If the stir bar flings off to one side or just rattles around, you may need to try again or adjust the distance slightly between the fan and the base of the snow globe.

GOING FURTHER

There are a lot of ways you can experiment with your Storm Globe and make it your own. Can you make a better lightning animation or more interesting LED lighting effects? Mount one on the end of a staff for an outstanding wizard costume effect? Maybe put waterproof lights inside the sphere? Connect it to a Lightning Detector circuit (*Make:* Volume 71, page 105) or trigger it with real-time lightning notifications from lightningmaps.org/apps?

I've noticed that the mica tends to eventually settle up against the walls of the globe. Easily fixed with an occasional shake and reset, but perhaps maybe a different stir bar would create enough turbulence to prevent it? (There's a good overview at makezine.com/go/stir-bar.)

Unfortunately we can't control speed directly from the Bluefruit, since the current draw of most fans is too high for it. However, you could certainly build a small circuit with a transistor to control the speed with the Bluefruit. This could be used to add a speed control knob or to slowly ramp up the stirrer.

I'd love to see what you can do with rheoscopic fluids and lighting effects; feel free to show off on Twitter and let me know @grajohnt! ◆

VIDEO RESOURCES
- My proof-of-concept video shows the swirling effect of a back-lit Kalliroscope: youtu.be/k0StfsKO8_U
- Video of the Atmospheric Movement exhibit at the Glasgow Science Centre that inspired this project: youtu.be/ LYTtutGOStU

```
def lightning(config):
    start_time = time.monotonic()
    last_update = start_time
    while time.monotonic() - start_time < config['duration']:
        if time.monotonic() - last_update > config['speed']:
            for _ in range(random.randint(1,8)):
                pixels.fill(0)
                pixels.fill(config['color'])
                time.sleep(0.02+(.001*random.randint(1,70)))
                pixels.fill(0)
                time.sleep(.01)
            time.sleep(5+random.randint(1,5))
            last_update = time.monotonic()
```

G

The Game of
Jeu Gruyère

Written and photographed by William Gurstelle

This simple woodworking project provides years of backyard fun, French style

The French are good at many things. French vintage wines are the standard to which the rest of the world's vintners aspire. French breads are *merveilleux*. So are French perfumes, designer clothing, and cheeses. And, I'd like to add one more thing to the list for admiring Francophiles: wooden games.

On a recent trip to Paris, I saw all sorts of Frenchfolk, young and old, playing in the spring sunshine with a variety of wooden toys and games. Many of them, like the now-popular cornhole game and giant Jenga, I had seen before. A few, like the Nok Hockey carrom board and wooden skittles, brought back memories from a long-ago past. But I saw one that seemed, at least to me, brand new and a lot of fun. The French call it Jeu Gruyère (or Jeu du Gruyère), which translates, more or less, to the "Swiss Cheese Game." (Figure Ⓐ)

Jeu Gruyère is like an enhanced, three-dimensional version of Chutes and Ladders mixed with Milton Bradley's Operation game. But it's more than that: when you add the ability to control the ball carrier using a pulley system, you get a game that's a challenge to both mind and body. The game consists of an inclined wooden board with holes placed at intervals and a wooden ball carrier that the facile Gruyèrist glides over the board via a system of ropes and pulleys that works just like a two-axis X-Y plotter. If the player is good, he or she can guide the ball to the top of the board without the ball falling through any of the Swiss cheese holes. If the player is not so good, well, then expect a fair amount of time watching the ball fall to the ground.

The game builder chooses the size and spacing of the holes in the board. Lots of holes, closely spaced, and a small-diameter ball create a tough challenge: mastering a Jeu Gruyère board like that takes a cool head and a keen eye and some practice. But for those a bit less dexterous and steady-handed, a board with wider paths and smaller holes makes for a perhaps more satisfying experience. In addition, you can vary the experience by using different shaped ball carriers, different sized balls, and longer or shorter rope handles.

TIME REQUIRED:
4–6 Hours

DIFFICULTY:
Easy

COST:
$40–$50

MATERIALS
- » **Plywood, ¾" thick, 48"×28"** Get the smoothest you can find; Baltic birch is a good choice. If you use rougher grades, make sure one side is sanded and filled.
- » **Pine or fir lumber, nominal 2×2, about 24' total length** Cut list: 58" legs (4), 18" leg brace (1), and 28" top brace (1)
- » **Board, 1" thick, 3¾"×3¾" square**
- » **Rubber or wooden ball, 1" diameter**
- » **Welded steel rings, 2" diameter (2)** for the handles. You can also use wooden balls for handles.
- » **Steel chain, #3 size, 10" lengths (2)**
- » **Deck screws: 2" (12) and 2½" (4)**
- » **Screw eyes, #12 (6)**
- » **Cup hooks, ⅞" (2)**
- » **Fixed pulleys, ¾" (2)**
- » **Paracord, ⅛" diameter, 12' total length**
- » **T-hinges, 3" (2)** with mounting screws
- » **Paint and primer**

TOOLS
Having the right tools for making the holes plays an important role in how safe and successful you will be in building this project. Here's what you'll need:
- » **Corded electric drill**
- » **Driver bit** for your deck screws
- » **Tape measure**
- » **Needlenose pliers (2)**
- » **Sandpaper, medium grit**
- » **Marking pencil**
- » **Hole saws: 3⅝" and 2" diameters**
- » **Safety glasses and gloves**

WILLIAM GURSTELLE's book series *Remaking History*, based on his *Make:* column of the same name, is available in the Maker Shed, makershed.com.

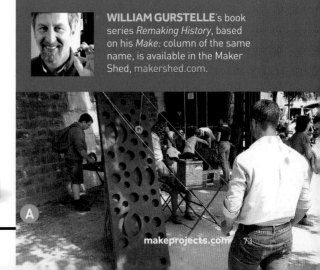

Ⓐ

DRILLING REALLY BIG HOLES

B

C

D

You have great latitude in making the Gruyère board's holes. While I chose to make all the holes on my board the same diameter, there's no reason you can't vary the diameters by using different sized hole saws. Be aware that bigger holes are harder to cut, and you'll need a fairly powerful electric drill in order to make them. I eschewed my 18V cordless in favor of a larger, more powerful corded drill (Figure B). While it is possible to drill large-diameter holes using a cordless drill in ¾" plywood, it is neither easy nor fun to do so. More power makes the job go better.

Further, the choice of the hole saw itself is an important one. A hole saw is basically a ring of sharp teeth surrounding a pilot drill bit in the center. The pilot drill bit keeps the hole saw from wandering around at the start of the cut. Since you're cutting through ¾" plywood, the hole saw's "cup" needs to be at least that deep. There are other methods of cutting holes in materials, such as circle cutters, which have the advantage of being adjustable, but I doubt the average circle cutter could handle the job of cutting so many large holes in plywood this thick.

The right tool for this job is a bi-metal hole saw Figure C). Hardware store shelves are covered with a variety of sizes of hole saws. You'll probably see two kinds of hole saws: carbon steel hole saws and the more expensive bi-metal hole saws. In the long run, you'll probably be a lot happier with the bi-metals. Bi-metal saws have teeth that are fabricated from two different types of steel. The outside surface of the teeth is made from hardened alloy that cuts easily through the plywood. But the inner surface of the blade is

made from springy, more ductile steel that allows smoother, easier cutting and a much longer lasting blade. A quality bi-metal hole saw will bore easily through materials such as plastic and wood and can handle metals as well.

The other thing to know about drilling holes is that good hole saws use an arbor, sometimes called a mandrel (Figure D), to attach the saw to the drill. Inexpensive hole saws have a rod extending from the back of the saw that goes directly into your drill's chuck. That might be fine for small-diameter holes or even a larger one if you're only drilling one or two. But if you're drilling many large-diameter holes, it's hard-to-impossible to apply enough torque to keep the drill from slipping in the chuck. So, hole saw manufacturers use a mandrel to connect the hole saw to the drill. The mandrel goes in between the saw and the drill. The mandrel attaches to the drill via its shaft. The hole saw itself attaches to the mandrel via a screw collar, but the turning torque is transmitted by two heavy duty drive pins

CAUTION: Take great care when drilling large-diameter holes. When the teeth of the hole saw contact the plywood surface, the saw applies a great amount of torque. This can result in the board itself spinning if it's not properly secured. Apply only slow, gentle downward force, making sure the drill is held perpendicular to the surface of the wood. Not doing so will cause the drill to wrench in your hands and that could cause injury. Go slow, wear safety equipment including glasses and gloves, and be mindful while drilling each and every hole!

BUILD YOUR JEU GRUYÈRE

1. Make the ball carrier by drilling a 2" hole in the center of the 3¾"×3¾" square board (Figure **E**).

2. Use four 2"-long deck screws to attach the top brace to the top edge of the plywood sheet (Figure **F**).

3. Use four 2" deck screws to attach each of the front legs to the unsanded side of the plywood.

4. Cut the Swiss cheese holes in the plywood (Figure **G**). See "Drilling Really Big Holes," facing page. I suggest starting out with a moderate number of holes — not too many and not too few. Admittedly, this is vague, but make your decision by assessing the coordination and ability of those who will be playing. Will the players be young children? Video gamers with great eye-hand coordination? People drinking beer? Space the holes accordingly. You can always add more holes later, but as far as taking them away, *non*, as the French say.

Make sure you don't cut holes too close to the outboard edges because you'll run into the 2×2s you attached to the back of the board in step 1. Sand the holes smooth with medium sandpaper.

5. Use the screws that come with the T-hinge to attach the back legs to the top brace. Then, use your 2½" deck screws to attach the bottom brace to the legs, about 16" up from the bottom (Figure **H**). Note that the back legs are located a bit inside the front legs. This allows the game board to fold flat for storage.

6. Attach the screw eyes to the front and back legs, 28" from the bottom of each leg. Then, connect the #3 chain to the screw eyes (Figure **I**). If necessary, you can adjust the length of the chain to provide the board inclination that works best for you. Use two needlenose pliers to open the chain links on the ends of each piece of chain. Insert the open links onto the screw eyes, and then use the pliers to reclose the links.

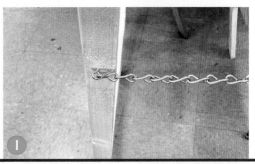

7. Prime and paint the game surface as you wish. Let it dry.

8. Next, drill pilot holes in the upper corners of the game board and screw in the cup hooks. Mount the fixed end of the pulleys on the cup hooks (Figure **J**).

9. Cut the paracord in two pieces, each 6' long. Attach the two remaining screw eyes to the sides of the ball carrier as shown in Figure **K**. Tie one end of each cord to each eye on the ball carrier. Then loop the cords through the pulleys and tie the remaining ends to the welded rings.

A CHEESY GOOD TIME

That's it! You're ready to play Jeu Gruyère and enjoy a bit of quality time with your friends and family. *Allez, on va jouer!*

- Place the game on flat ground and make sure it won't tip over.
- Place the rubber ball in the ball carrier and position it at the bottom of the board.
- Use the ropes to manipulate the ball and see how quickly you can get to the top without losing the ball into the holes. *Bon chance!* ⊘

LEVEL UP YOUR GRUYÈRE GAME

Once you've got the hang of Jeu Gruyère, there are all sorts of variants to try:

- **SCORING** — Mark off the board in four sections from bottom (easy) to top (hard) and assign 1–4 points, respectively. Assign extra points for holes that are hardest to reach. For skilled players, dropping in zone 1 means you lose, you're out (Figure **L**)!

- **MAKE BIGGER, ODD-SHAPED HOLES** — with narrower, more challenging paths between them

- **DRAW COURSES ON THE BOARD** — easy or difficult, see who can follow them fastest

- **TWO-PLAYER TEAMS** — each take one handle, and work together to guide the ball

- **LOOP THE HANDLES AROUND YOUR KNEES** — for a more challenging game called Piège à Genoux (Knee Trap) (Figure **M**)

- **DOUBLE-SIDED GRUYÈRE BOARD** — Both sides can be played solo at the same time, or players can race each other. A fabric stretcher catches dropped balls (Figure **N**).

- **BALL RETURN** — Build a shallow box behind the board to capture the ball and return it to the bottom (Figure **O**). If you're really ambitious, you can partition the box to deliver the ball to individual ball returns, assigning each a different score (Figure **P**).

—*Keith Hammond*

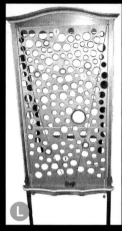

L

wellouej.com

M

alortujou.com

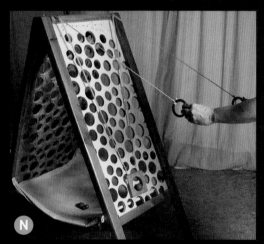

N

tujoues.fr

O

woodbotherer.co.nz

P

jeuxdautrefois.free.fr

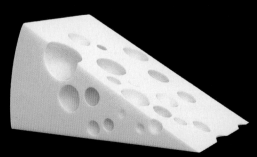

Adobe Stock - Anna

DIY Atlas of Space
Going to the Moon and Mars?
You'll need maps. Beautiful maps.

Written and photographed by Eleanor Lutz

ELEANOR LUTZ (eleanorlutz.com)
is an award-winning information
designer in Seattle, Washington.
She gets excited about maps, emoji,
and indoor plants, and she blogs at
tabletopwhale.com.

For two years I've been working on a collection of ten maps of planets, moons, and outer space. To name a few, I've made an animated map of the seasons on Earth, a map of Mars geology, and a map of everything in the solar system bigger than 10km. I call it my Atlas of Space.

Last summer I began sharing each map, along with the open-source Python code and detailed tutorials for re-creating the design. All of the astronomy data comes from publicly available sources like NASA, USGS, and the International Astronomical Union (IAU), so I thought this would be the perfect project for writing design tutorials (which I've been meaning to do for a while).

THE FINISHED TUTORIALS NOW COVER:
- Working with Digital Elevation Models (DEMs) in Bash and Python
- Working with ESRI shapefiles in Python
- Using the NASA HORIZONS orbital mechanics server and scraping internet data
- Working with NASA image data
- Combining many datasets into one map
- Color palettes and style design
- Decorative illustrations and painting in Photoshop
- Plotting with symbols and different languages in Matplotlib
- Editing Python outputs in Illustrator
- Using the IAU Gazetteer of Planetary Nomenclature
- Map projections in Python Cartopy
- Mapping constellations using star catalog data.

I originally learned Python for my grad school research, which involved processing video recordings of mosquito behavior. I really liked the language, and now I use Python extensively for data management and graphic design. I also love teaching Python — I volunteer for Software Carpentry (software-carpentry.org), and I taught Data Science for Biologists at UW last year with Bing Brunton and Kam Harris (github.com/eleanorlutz/Data_science_for_biologists_2019).

So if you're interested in beginner-friendly explanations of Python cartography, web scraping, or using the NASA orbital mechanics server, this is the project for you.

Software used includes Python 3.7.1, GDAL

TIME REQUIRED:
A Weekend

DIFFICULTY:
Intermediate

COST:
$0

TOOLS
» **Computer running Python 3.7.1 and GDAL 2.4.1,** both free and open source
» **Image editing software, raster and vector** I used Adobe Photoshop CC 2019 and Illustrator CC 2019, but you can also use the free open source programs Gimp and Inkscape.

2.4.1, NASA HORIZONS, Adobe Illustrator and Photoshop. Python dependencies include `matplotlib`, `numpy`, `pandas`, `os`, `cartopy`, `json`, `osgeo`, `math`, `scipy`, and `jupyter`.

Each of the following tutorials includes special instructions for beginners, graphic design advice, and all of my open source code, ready to run. Find them at at github.com/eleanorlutz?tab=repositories. You can do this.

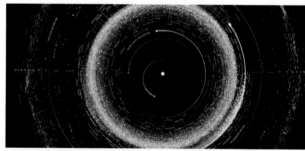

SOLAR SYSTEM ORBITAL MAP

This tutorial shows how to map the orbits of all the planets and more than 18,000 asteroids. This includes everything we know of that's over 10km in diameter — about 10,000 asteroids — plus 8,000 randomized 'roids of unknown size. I used the NASA JPL Small-Body Database Search Engine to make a list of all known asteroids and comets in the solar system, and pulled the planets, moons, and trans-Neptunian objects from other NASA datasets. All this data is public but it's in several different databases so I had to do a decent amount of data cleaning. The map shows each asteroid at its exact position on New Year's Eve 1999.

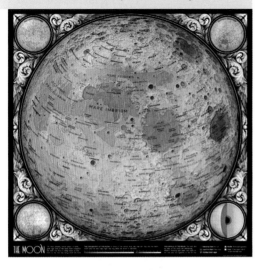

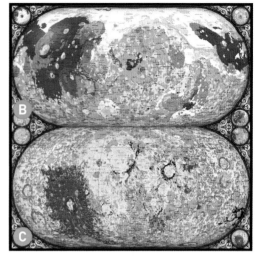

TOPOGRAPHIC MAPS OF THE MOON AND PLANETS

This shows how to make topo maps of planets and moons using open-source data from the USGS, IAU, and NASA. I wanted to map each of the rocky planets in the same style: Mercury, Mars, Venus, and the Moon. To begin, I used GDAL to convert the DEM files to orthographic projection, then downsampled these and created hillshade and slope maps for each hemisphere. Next, I made five Python plots with the contour fills, contour lines, text labels, and two types of gridlines (Figure A). Finally, I used **matplotlib gridspec** so that each of my subplots occupy the exact pixel locations inside my decorative border.

I also designed a cutaway diagram showing the interior layers of each planet. These were tricky because some layers were so thin that they were virtually invisible. To show even the thinnest layers, I designed an adjusted diagram where every layer has a minimum visible thickness.

This project was heavier on the illustration side than my asteroid map. I had a lot of fun designing the scrollwork, and I also designed Photoshop overlays to add a 3D effect to the globes and the cutout core diagrams.

GEOLOGIC MAPS OF THE MOON AND MARS

For this I used open-source data from the USGS, IAU, and NASA, but I added more topographic and label data, redesigned the visual style. The Mars map (Figure B) is an artistic rendition of the USGS map (pubs.usgs.gov/sim/3292). I also edited the key for a more general audience. For example, my abbreviated definition for a caldera rim is, *The rim of an empty magma chamber left behind after a volcanic eruption.* The original description was, *Ovoid scarp, outlines single or multiple coalesced partial to fully enclosed depression(s); volcanic collapse, related to effusive and possibly explosive eruptions.* Right!

I wanted to accurately show how much of the planet was made up of each geologic formation, so I decided to use an Eckert IV equal-area projection. This type of map distorts object outlines, but it preserves the relative area of shapes across the globe. Eckert IV is not great for visualizing the polar regions, so I added four inset maps to the corners to show each hemisphere of Mars.

The Moon map (Figure C) was much more difficult. The geologic data was split into six different datasets, each with unique labels and

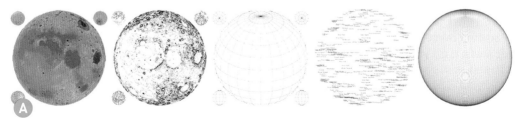

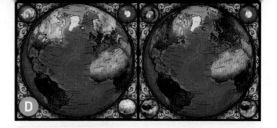

some with different data formats. Where geologic categories were described differently, like "Basin Material, Rugged" vs. "Material of Rugged Basin Terrain," I combined closely related terms into a single color. Also, the geologic timescales weren't precise, so I decided to omit timescale data. The map features are colored by geologic category (craters, basins, etc.) and not by age.

ANIMATED EARTH

This uses open-source code and data from NASA, USGS, and Natural Earth. NASA publishes many beautiful Earth datasets at monthly time scales, and this GIF uses one frame per month to show the fluctuating seasons, focused mainly on data about vegetation and Arctic sea ice (Figure **D**).

ANIMATED JUPITER

Just for fun, this animated GIF illustrates the storms on Jupiter (Figure **E**). Created mostly in Photoshop, based on photos and video published from the Cassini spacecraft in 2000 and 2006.

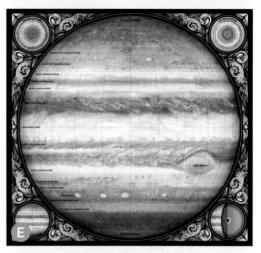

THE NIGHT SKY

Western constellations — Shows every star visible from Earth, plus the brightest galaxies, nebulae, and star clusters from W.H. Finlay's *Concise Catalog of Deep-sky Objects*. I illustrated the familiar Western star patterns — or *asterisms* — in blue and gold, and the scientific constellation boundaries in red (Figure **F**), using open-source data from Stellarium and the HYG Database.
World constellations — Includes the animals, people, and objects imagined in the sky by more than 30 civilizations including ancient Egyptian, Arabic, Korean, Chinese, Japanese, Dakota, Hawaiian, and Mongolian cultures (Figure **G**).

DIY MAP DESIGN

All of these tutorials include sections on how to design your maps in Python, Illustrator, and Photoshop. I've also shared a piece of the scrollwork illustration file that includes the original layers as reference, if you want to use a similar style in your own projects.

I love working with astronomy data. I'm sure I'll make more maps in the future, but I had to take a break to finish my last PhD research project and my current fellowship at *The New York Times*. ◗

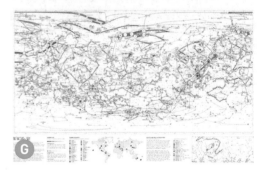

GET THESE MAPS:
- Wallpapers: tabletopwhale.com/2019/07/08/an-animated-map-of-earth.html
- Buy prints: redbubble.com/people/EleanorLutz/explore

Written and photographed by Jane Stewart

Knot the Child You're Looking For

This is the way — to tie your own macramé Baby Yoda

Last November the world met a precious little creature officially known as the Child, in the new TV series *The Mandalorian*. The internet instantly named him Baby Yoda. I was already planning to do more *Star Wars* macramé (I'd previously made an R2-D2, lightsaber, and Death Star) and Baby Yoda was so cute I just had to try.

Seven months later and I've made over 50 macramé dolls of him, become a Facebook moderator on a fan page, and even founded my own FB group chronicling my mini Baby Yodas.

My first design was based on the show, but with a sparkly coat. Here's how I made him.

MAKE YOUR BABY YODA

First you'll tie the structural threads, then you'll tie rows of knots onto these, following the chart (Figure Ⓐ). Using standard embroidery thread, this pattern makes a Baby Yoda about 4cm tall; use bigger yarn for bigger babies. For complete step-by-step photos, visit the project page online at makezine.com/go/macrame-baby-yoda.

SET UP STRUCTURAL THREADS

Choose your 5 colors. I used 6-strand embroidery floss in green, and sparkly metallic 3-ply crochet yarn in black, white, gold, and copper.

Measure four 50cm lengths of green thread and the same in gold. Bend in the middle, and tie them together with a simple overhand loop.

Tie another overhand knot in the middle and splay out the threads: 4 green then 8 gold then 4 green (Figure **B**). These are the structural threads. They're mostly unseen, being the threads that everything else is tied around .

DOUBLE HALF HITCH KNOT

You'll tie a few hundred of these:

C Cross the green thread over the gold.
D Loop round the back and over the top.
E Repeat, running the free end under first loop.
F Pull tight to finish.

MAKE THE HEAD

Leave at least 5cm green thread on both ends of the 16-knot rows (rows 1, 5, 9, 13, 17, and 21). These will become part of baby's ears.

Row 1: In green, make the double half hitch knot over every structural thread. This row is 16 knots long. Trim ends to 5cm or longer.

Row 2: Miss out the first structural thread, and make 14 knots in green. Trim short (1cm–2cm).

Row 3: Miss out 2 more structural threads, make a row 10 knots long, and trim short.

Row 4: Miss out 2 more structural threads, make a row 6 knots long in green, and trim short (Figure **G** on the following page).

Rows 5–16: Now you'll repeat rows 1–4 three more times. Begin row 5 just like row 1, tying onto the first structural thread in green. The chart shows the rows in a schematic way; really you're tying this row snug against all the previous rows. This creates the curvature of the head.

As you work through row 5 (Figure **H**), tuck the ends of the previous rows behind the emerging fabric. Keep following the pattern until you've completed row 16.

Row 17: It's time to start the eyes: tie on white and black (Figure **I**). Working L–R, tie 4 knots in green, 1 black, 1 white, 4 green, 1 black, 1 white, 4 green (Figure **J**).

Row 18: Miss out first structural thread and tie on 2 green knots, 4 black, 2 green, 4 black, 2 green.

Row 19: Miss 2 more structural threads and tie 1 green, 2 black, 4 green, 2 black, 1 green. The eyes are complete.

TIME REQUIRED:
4–5 Hours
DIFFICULTY:
Intermediate
COST:
$5–$20

MATERIALS
» **Cotton embroidery floss, thread, yarn, or macramé cord in 5 colors** pale green, black, white, and two brownish ones for the robe

TOOLS
» **Scissors**

JANE STEWART is a lifelong crafter. For the last few years she has been tying many, many knots in things. She created the Macramé Periodic Table of the Elements featured in *Make:* Volume 67 and exhibited around the UK in 2019.

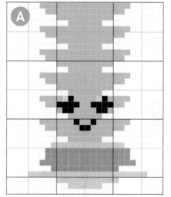

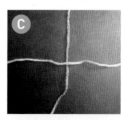

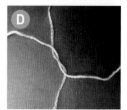

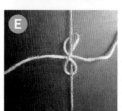

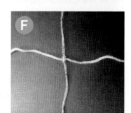

Row 20: Miss 2 more structural, tie 6 green knots.
Row 21: Make the mouth: 6 green, 1 black, 2 green — then knot 2 black on row 22 — 1 black back on row 21, then 6 green (Figure Ⓚ). The mouth looks better when it's continuous like this.
Row 22: Miss 1 structural thread, tie 6 green, miss the 2 black you just did, tie 6 more green.
Row 23: Miss 2 more structural threads, tie 10 green knots.
Row 24: Miss 2 more structural threads, tie 6 green knots. The head is complete.

Tie structural threads 1 and 2 front to back, and tie 15 and 16 the same way (Figure Ⓛ).

MAKE THE EARS

Those long ends from your 16-knot rows will now serve as structural threads for the ear. Separate them into 3 from the front of the head, and 3 from the back. For row 1 of the ear, tie 5 green knots, from bottom to top, alternating like so: 1 back, 2 front, 3 back, 4 front, 5 and 6 together (Figure Ⓜ).

Tie row 2 the same way, but in the reverse direction. Row 3 changes direction again, 4 knots, this time gathering structural threads 4, 5, and 6 inside the final double half hitch knot at the top (Figure Ⓝ). Row 4, again changes direction. Keep knotting back and forth, taking in a structural thread each time you reach the top of the ear. Eventually all will be gathered inside your final knot. The ear is done (Figure O).

Trim off the excess thread on the end of the ear. Repeat for the second ear.

MAKE THE COLLAR/SHOULDERS
Untie the overhand knots in the structural threads.
Row 25: Starting at the front, use 4 gold threads together, knotting over 2 structural threads at a time. Knot all the way around the base of the head; the last knot will be gold tied around itself. Leave these long; they're now structural threads!
Row 26: Use 2 copper threads together, knotting over 2 structural threads, to knot the first copper row on the body, starting and ending in the center of the back. Leave long, at least 5cm (Figure P).

BEGIN THE ARMS
Plait the green structural threads 1–3 to make an arm, long enough to reach up to the eye and back to the shoulder (Figure Q). Tie the end of the arm back to the shoulder, using the loose fourth green structural thread to pull the end of the plait inside the body. For the second arm, repeat on the opposite side (Figure R).

Trim excess green length from below the shoulders. Tuck the loose green ends inside the head. Then tie the bases of the arms together, between the front and back structural threads.

MAKE THE BODY
Row 27: Begin the second row of copper in the middle of the back, 2 threads together tied over 2 structural threads. Continue all the way round, beneath the arms, 15 total knots (Figure S).
Row 28: Third row of copper, 17 knots.
Row 29: Fourth row of copper; miss out two structural threads at the beginning and end of the row, 14 knots. Leave long to become new structural threads.

MAKE THE BASE
Row 30: Starting front and center, begin the first row of the base, using 2 gold threads, knotting around 2 structural threads, 21 knots (Figure T).
Row 31: Tie the second row on the base, knotting around 4 structural threads.
Row 32: Tie the third row, knotting around 8

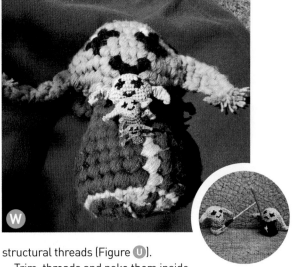

structural threads (Figure U).

Trim threads and poke them inside the body to make a flat base (Figure V).

MAKE THE SLEEVES
Knot copper thread around the arm 8 times. You can knot in the same direction if you want, or alternate directions to get a straight ridge on one side of the arm. Tie the copper ends to the body and poke the thread inside the body.

Make the second sleeve the same as the first. It's finished! 🤍 🤍 🤍

THE KID'S COMING WITH ME
Baby Yoda makes a keen gift for fellow Mando nerds. You can make different sizes (Figure W) simply by varying the thread (or yarn) size. Variations on my first design involve different colors, thread thicknesses, expressions on his little face, and naked baby versions where you can see his little feet.

I've also made several macramé creatures that Baby Yoda has been compared to (Figure X). Most similar is the *Toy Story* alien, much like Baby Yoda but with upturned ears, three eyes, and antenna. Pikachu and Stitch also have very similar body designs. The Minion Stewart is much simpler. See them all at instagram. com/macramebrainjane and facebook.com/ groups/525098275035176.

Take care of this little one. Or maybe, it'll take care of you. ❷

Roth's Orbitron

Written and photographed by Bob Knetzger

Making the Hot Wheels car ... that never was!

Ed Roth was a huge influence on 1960s kids, and I was no exception. We were crazy for all his kooky creations: weirdo T-shirts and decals, plastic Rat Fink and car models, slot car magazines and comic books, and Hot Wheels cars. It all started with Roth's wild custom hot rods, seen at car shows and in magazines. Revell licensed his car designs for model kits and created the moniker Ed "Big Daddy" Roth (he had five kids). Testors sold "Roth Custom Finishes" paints in colors named after his cars. Mattel re-created his Beatnik Bandit as one of the first Hot Wheels cars. Kids eagerly anticipated Roth's next car and gobbled it all up.

But in 1964 when Roth launched Orbitron, his fan base of little finksters (as he called them) turned fickle. The fad had faded: The Beatles had captured the world's attention and kids turned to rock instead of 'rods. Disenchanted, Roth switched gears to making motorcycles and launched *Choppers* magazine. He sold off his cars and the Orbitron disappeared, never to be produced as a toy or model.

Until now!

I've always wanted an Orbitron, so I had to make one myself. Here's how I kit-bashed and carved my own 1/64-scale Hot Wheels version. Use these tools, tips, and techniques to make your own custom creations!

1. KIT-BASHING FOR PARTS

I assembled lots of reference photos to plan my Orbitron. First up: chassis and wheels. I culled some "Real Rider" rubber tires and wheels from new and vintage HW cars, using different cars for the front tires and rear slicks (Figure ).

Drill out the rivet heads to separate the body and chassis, then carefully pry out the axles (see *Make:* Vol. 65, "Hot Mods" for details, makezine. com/projects/hot-mods). I had to make new, wider axles from 0.030" wire to get the wheelbase and stance right. I cut down the die cast chassis to the proper proportions and relocated the axles (Figure ).

One of the HW cars had a suitable chromed miniature plastic engine and a third car — Speed Racer's Mach 5 (Figure ) — was cannibalized for the interior.

BOB KNETZGER is a designer/inventor/musician whose award-winning toys have been featured on *The Tonight Show*, *Nightline*, and *Good Morning America*. He is the author of *Make: Fun!*, available at makershed.com and fine bookstores.

Foam sculpting tools: modeler's saw, hand files, dental picks, Dremel with bits, emery boards, and magnifier

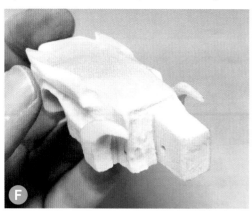

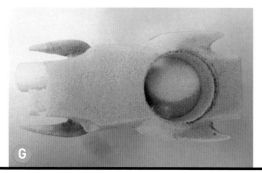

2. ONTOGENY RECAPITULATES PHYLOGENY — IN URETHANE!

Roth made his fiberglass car bodies without molds or detailed plans. He created rough forms right on the finished chassis with wood, chicken wire, and globs of plaster. Next, he laboriously carved the finished shapes by hand. After he covered it all with fiberglass, he'd knock out all the plaster from underneath the finished body. Each car was a one-of-a-kind!

I made my Orbitron in a similar way. With the chassis and wheels as a starting point I cut a small block of 30-pound-density rigid urethane foam. Industrial designers use this uniform, no-grain material to quickly create forms and prototypes because it's very easy to sculpt with hand tools. I formed the body shape by hand, then sealed and painted the foam to make a one-of-a-kind Hot Wheels body.

> **TIP:** Even in this small size, there will be lots of super-fine urethane dust — you'll want to wear a painter's mask and vacuum up as you go.

Begin by sketching the basic shapes of the top, front, rear, and side views right onto the foam (Figure **D**). I slightly exaggerated the proportions (which is often done in Hot Wheels) to make a "cuter" version in this miniature scale.

To start the sculpt, cut the rough shape using a band saw. A modeler's saw (Figure **E**) is great for precise and small cuts. Use a Dremel with a sanding drum to make the wheel well cutouts.

The front fenders took extra time and patience with lots of hand filing (Figure **F**). A tiny ball cutter in the Dremel can hog out odd-shaped cavities and the headlight recesses.

Go slowly and check the shape against the reference photos. Test-fit the body to the chassis and parts. As the shape gets more refined, use smaller files and finally just sandpaper.

> **TIP:** Manicurist's emery boards are very useful for sanding in tight places on miniatures. You can also make your own sanding tools by mounting sandpaper on boards and wrapping fine grit sandpaper around small dowels.

I used a ¾" end mill to make a flat-bottomed circular pit to receive the interior (Figure **G**), then

I filed down the plastic interior part to fit neatly inside. A small square of styrene became the miniature TV in the dashboard (Figure **H**).

The Dremel and a cylindrical bit made short work of hogging out the engine compartment. Careful hand filing squared things up.

> **TIP:** If you break off a chunk of foam you can reattach it with cyanoacrylate (CA) glue, but because of the foam's texture you'll have to first apply a layer of CA to each surface, let it cure to a thin film, then bond the two parts together. If you need to add material, use Plastic Wood filler. It will bond tightly to the foam and it isn't too hard. Don't use harder epoxies or polyester like Bondo — it's impossible to file or sand along with the much softer foam!

Materials: pin striping paint, MEK solvent, Plastic Wood, glazing putty, foam primer, spray lacquer.

3. FABRICATE CUSTOM PARTS

The hood was scratch built from various thickness of styrene, solvent bonded with MEK and trimmed to fit (Figure **I**). The bubble top was vacuum-formed out of 0.010" clear styrene using a marble as a mold held in place by modeling clay (Figure **J**, following page). A thin collar on the marble creates a flat rim on the formed bubble at just the right diameter to fit the car. (Later I'll use that collar on the bubble as trim.)

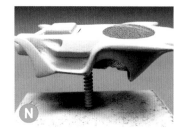

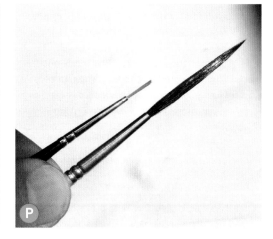

I turned and beveled a piece of styrene rod on the lathe to make the cylindrical nose. I milled a flat-bottom hole in the body with a ⅜" end mill and CA'ed the nose in place (Figure **K**).

4. FINISH AND PAINT

Once the form is well defined, seal it with fast-hardening glazing putty. Using a tiny spatula cut from a plastic lid, trowel on a skim layer of putty. Immediately scrape it off, leaving putty in just the pores of the foam texture.

Paint the body with white primer (Figure **L**) to see the shape better and identify areas to re-putty and sand (Figure **M**). It will take a few iterations of layers of primer and putty to get a smooth, finished surface (Figure **N**).

After a final coat of primer on the body, bubble trim ring, and hood, spray on the color: a medium blue metal flake lacquer (Figure **O**). Oh, how I wish I could get some 1960s Testor's Ed "Big Daddy" Roth Custom Finish in Orbitron Blue!

The last step is the trickiest: pin striping. Real pin striping is hard enough, but in ¹⁄₆₄ scale? Yikes! You'll need a miniature pin striper's brush. Get a 20/0 "rigger" brush with extra long bristles (Figure **P**) to hold more paint for brushing l-o-n-g lines in one pass. Just as in full-size pin striping, stroke the brush across the palette to load up some paint, then drag the brush held at a shallow angle (Figures **Q** and **R**). Use a smooth, continuous motion and don't swivel your wrist.

A ONE-OF-A-KIND CUSTOM!

Here's the assembled Orbitron with some other Roth Hot Wheels: Mysterion, Road Agent, Outlaw, and Beatnik Bandit (Figure **S**).

To display this vintage "toy-that-never-was," I created a blister card package with retro graphics (Figure **T**). Swell! ●

TIP: Practice pin striping first on some other old Hot Wheels cars until you can make thin, even lines. Good luck!

RESTORED TO GLORY

What ever happened to the real Orbitron? It was sold to someone in Texas and lost for over 20 years. Beau Boeckmann of Galpin Auto Sports heard a rumor it was spotted in Juarez, Mexico, stripped to a hulk, and left on the street. It was rescued and longingly restored by a team of craftsmen, many of whom worked on the car originally. (GAS now has a show, *Car Kings*, on Discovery Channel.) I saw the restored Orbitron, along with Roth's Rotar, Mysterion, and Tweedy Pie, on display at Galpin Ford in Van Nuys, California.
» Orbitron Rides Again! youtube.com/watch?v=5TS-o2TLpMA
» Orbitron at the GAS collection: galpinautosports.com/collection-home

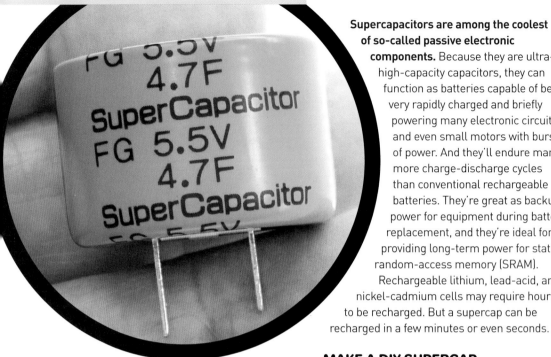

Badder Than Batteries

Getting to know the super fast-charging supercapacitor

Written and photographed by Forrest M. Mims III

FORREST M. MIMS III
(forrestmims.org) an amateur scientist and Rolex Award winner, was named by *Discover* magazine as one of the "50 Best Brains in Science." His books have sold more than 7 million copies.

Supercapacitors are among the coolest of so-called passive electronic components. Because they are ultra-high-capacity capacitors, they can function as batteries capable of being very rapidly charged and briefly powering many electronic circuits and even small motors with bursts of power. And they'll endure many more charge-discharge cycles than conventional rechargeable batteries. They're great as backup power for equipment during battery replacement, and they're ideal for providing long-term power for static random-access memory (SRAM). Rechargeable lithium, lead-acid, and nickel-cadmium cells may require hours to be recharged. But a supercap can be recharged in a few minutes or even seconds.

MAKE A DIY SUPERCAP

While commercial supcapitors are best for DIY projects, you can learn about their performance by making one yourself. A very simple supercap can be made with activated carbon, aluminum foil, and lemon juice.

Figure Ⓐ shows a DIY supercap made from a single Tetra Whisper Filter Cartridge meant for aquariums. Open the activated carbon pack, pour the contents into one of the filter bags, and seal the bag with a stapler. Next, cut two pieces of heavy-duty aluminum foil slightly smaller than the filter bag. Then sandwich the filter bag between the two sheets of foil, making sure the opposite foil sides do not touch.

Place the completed supercap in a shallow container with a weight on top to hold it in place. Or secure the layers together with tape, again making sure the opposing foil sides don't touch. Finally, pour some lemon juice over the supercap to provide an electrolyte.

The DIY supercap should be charged with 1.5 volts or less to avoid decomposing the lemon juice. The one shown in Figure A was charged to 1.31 volts from a single AA cell. Figure Ⓑ shows a charge-discharge cycle of this supercap charged to 0.9 volt and recorded by a 16-bit analog data logger (an Onset HOBO 4-channel). The most obvious feature is the nearly instantaneous charge time, which is far faster than typical conventional capacitors.

You can charge a DIY supercap to a higher voltage by stacking multiple layers. In *Science and Communication Circuits and Projects* (sparkfun.com/products/11132), one of my RadioShack Engineer's Mini-Notebooks, I describe how to make a multi-cell supercap by stacking 5 layers of premade activated carbon sheets purchased at an aquarium store. The sheets are separated by slightly larger sheets of paper towel. Bare, copper-clad PC boards are placed on opposite sides of this 5-layer sandwich, which can be charged to 3 volts or more. This DIY supercap will power a red LED for several minutes or longer.

COMMERCIAL SUPERCAPS

You'll need to use a commercial supercap for practical applications. Many are available, and their prices have dropped significantly since they were first introduced.

Page 92 shows an NEC supercapacitor that can be charged to 5.5 volts. It has an incredible capacitance of 4.7 farads! This means you can charge this supercap from an active USB port and use it to power many low-power circuits.

Figure ⓒ shows the charge-discharge cycle of this supercap when connected across a 470-ohm resistor. While the discharge is faster than might be desired, the charge time is only about 10 seconds.

GOING FURTHER

You can perform many experiments with DIY supercapacitors. Try stacking them to increase their capacity, or substituting the lemon juice with other electrolytes. How does the charge-discharge cycle change when vinegar is substituted? Or when salty water is used?

The electrolyte substrate used here is the filter bag provided with the activated carbon. But you can also use a paper towel or fabric.

Finally, you can experiment with the physical configuration of DIY supercaps. For example, you can roll your sandwich of foil/carbon/foil into a cylinder. Insert the cylinder into a pill bottle and pour in some electrolyte. ◗

TIME REQUIRED:
1–2 Hours

DIFFICULTY:
Easy

COST:
$10–$20

MATERIALS
» **Aluminum foil**
» **Activated carbon** Regent Aqua-Tech or Tetra Whisper Filter Cartridges, from aquarium store
» **Lemon juice** bottled or fresh
» **Clip leads (2)**
» **Tray**
» **Battery, 1.5V or less** AA, AAA, C, or D size

TOOLS
» **Stapler**
» **Voltmeter (optional)**

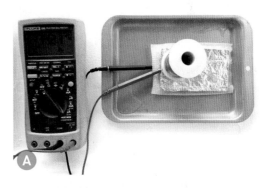

Ⓐ

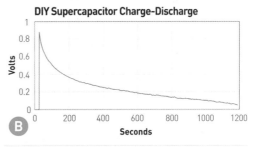

DIY Supercapacitor Charge-Discharge

Ⓑ

NEC Supercapacitor Charge-Discharge

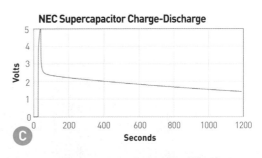

Ⓒ

Written by Mimi the Nerd

Mimi The Nerd as 2B
from the videogame
Nier: Automata, 2019.

MIMI THE NERD (@mimithenerd) is a makeup
artist, cosplayer, writer, horror and anime lover,
and mental health advocate in England, U.K.

Cosplaying While Black

Tips on getting started and ignoring the haters

Since I was young I have always enjoyed dressing up as different people, from celebs to anime and cartoon characters. I wasn't aware at the time that this was indeed called *cosplaying*. I had much to learn. Here's what I've gathered.

GETTING STARTED IN COSPLAY

Want to join the cosplay community but don't know where to start? It can all seem really overwhelming, but I'm here to tell you there's no need to panic.

My first tip is always to start small and use what you have: this is called **wardrobe cosplay**. For example, my Spinelli cosplay from *Recess* required only a leather jacket, red shirt, and an orange hat. The only thing I had to purchase was the orange hat which cost less than 5 quid ($6.50). And my Numbuh 3 from *Codename: Kids Next Door* required only a green jumper and straight black wig with bangs (Figure A), which I already had. To be a "cosplayer" you don't have to become the characters with super complex costumes. Every costume is valid, whether it's bought or made.

Halloween is a great time to experiment with different types of makeup and costumes. I used to be that person who was afraid of face paint because it is a whole different game compared to actual makeup cosmetics. The techniques are different, but I made it my mission to buy some paints and just experiment with characters that required full face and body paint, and now I can actually say I am good at it, and have made it my own (Figure B and Figure C on the following page). So my advice here would be: That thing that you're scared to start? Do it!

I became more serious with my cosplay in 2015 when I attended my first **cosplay convention** —

Full face paint achievement unlocked: Camilla from *A Bad Case of Stripes*, 2019

Linda Blacker, Mimi the Nerd, Adobe Stock - Vjom

where people who share the same interests in anime, manga, comics, TV shows (basically anything that creates a fan base and can be cosplayed) gather to hear talks, dress up as their favorite characters (Figure ), meet and greet the actors who play said admired characters, and buy merchandise of their favorite shows. MCM Comic Con is one of the biggest global cosplay convention companies with events in London, Birmingham, and Manchester.

It wasn't until 2018 that I took part in the **online cosplay community** to present my art of cosplaying, and from there have I have grown a supportive following on Instagram and Twitter. But of course, not everyone is supportive.

COSPLAYING WHILE BLACK

The lack of ethnic diversity in the cosplay industry stems from a shortage of black people playing these characters in manga, TV shows and films. Many black cosplay artists would agree there's a lack of representation and a stigma within the cosplay community for people of color. There is still a lot of racism, and black cosplayers are usually the target.

When I had first joined the cosplay community and started to get properly involved in the anime community, I must say I was naive and thought

Emily from *The Corpse Bride*, 2019

Gathering of the tribe: *Steven Universe* convention meetup, Mimi as Garnet at left.

Mimi the Nerd, Adobe Stock - Vjom

everyone would support each other and it would be full of rainbows and fairy dust. However, we still have a long way to go to be fully accepted in this space. Even though it's 2020, there's always a black cosplayer getting abused by a racist online and it does need to stop.

Any online personality knows there will always be negative comments and internet trolls who are not fans of their work. This is something I am all too familiar with. I have the support of my fans and followers, which I am grateful for, but I'm also exposed to nasty comments that include racist or ignorant thoughts. One comment that comes up a lot is when some people say I cannot cosplay a certain character because the character is not black. This is pure ignorance. Any cosplay is a person's take on a character, not an exact mirror image of the character. How someone wants to represent a character is completely up to them.

Creating cosplays is something that helps me keep my mind off things and a way I cope with my mental health. I will not let these comments get in the way of me continuing to be active in the cosplay community and I will forever use my platform to use my voice. These trolls cannot stop me, and you shouldn't let them stop you either. These are people hiding behind their computer screens. They're most likely jealous.

My journey as a cosplayer has been quite a rollercoaster, to be honest. I wasn't always confident, and I wasn't always vocal. But the more I knew I wanted cosplay to be more than just a hobby, the more I grew confidence. The more characters I did, the happier it made me, and I would always look forward to whatever I came up with next. It also helped cosplaying quite strong characters that I admire, because they ooze confidence.

THE YEAR OF COVID AND BLM

Covid-19 has caused 2020 comic conventions to be canceled, and cosplayers have been dealing with it in their own ways. Some are using this time to make bigger and better costumes so they can be ready for the next convention. Some are doing virtual conventions featuring special cosplayer guests, which is a super cool idea. Others, me included, are using this time to create as much content as they can for their socials.

The recent events taking place and the Black Lives Matter movement are not new, but they have sparked conversations on social media now more than ever. There has been an influx of support for black creatives and we're starting to get more recognition because of hashtags like #blerd and #28daysofblackcosplay, and challenges such as #amplifymelanatedvoices, founded by mental health therapist Alishia McCullough and activist Jessica Wilson, which invited users to spend a week promoting work by black creators rather than their own. The intention was to show solidarity and amplify black voices because black creatives are often overlooked in comparison to those who are white or white passing. When it comes to allies in the cosplay community, I feel like a lot of non-black people are still very quiet on these issues and don't really fight with us when things happen, which is a shame because you'd think as a community, we are one. I have hope that things will change for the better, but like I mentioned, there's still a long way to go.

Lastly, my most important tip is to just go for it, have fun! It's so important to not take yourself too seriously when cosplaying because that's when you'll be too in your head and stress out. The definition of cosplay is literally "costume play" so play around, have fun, and don't stress. At the end of the day, do what makes you happy and forget what anyone negative has to say because they don't matter. ◙

AMPLIFY MELANATED NERDS

I have quite a few cosplay faves but I'd love to highlight these individuals in particular as I don't think they get the recognition they truly deserve. Each is unique and puts a different spin to the characters they cosplay. Find them on Instagram. ◐

PETITE EBBY OF SAVANNAH, GEORGIA

@chibimagicalgirl
Yuna from the videogames *Final Fantasy X* and *X-2*, summoning monsters at Colossal Con 2018. "I've been a cosplayer for 9 years now. Throughout my cosplay journey, cosplaying has helped me so much: meeting new people, being an advocate in the cosplay community, showing and providing cosplay positivity, and inspiring people along the way."

STARDUST MEGU OF THE BRONX, NEW YORK

@stardust_megu
Athena (Saori Kido) from the manga/anime *Saint Seiya: Knights of the Zodiac*, 2019. "I made the whole cosplay myself (including the petticoat underneath)!" Satin dress and tulle petticoat, craft foam accessories, staff of EVA foam and PVC pipe, gems of cast resin. Winner, Judge's Choice, International Cosplay Day, Central Park, NYC.

ANDRIEN GBINIGIE OF MONTREAL, QUEBEC

@escoblades
Black Panther, Green Lantern, Superman, and Shazam, all 2018. "'You get to decide what kind of king you are going to be.' The character that started it all for me: Black Panther."

Ken Austin @shiningeternity, Crissy M Photography @crissymphotos, Michaëlle Charette, michaellecharette.com (GL, Superman,

SHAIHIEM TYGEE KING OF NORTH CHARLESTON, SOUTH CAROLINA

@tgyeesensei
"If you wanna stop this, then stand up! Because I've just got one thing to say to you! Never forget who you want to become!" — Shoto Todoroki, from *My Hero Academia*, 2020.

DORASAE ROSARIO OF CINCINNATI, OHIO

@akakioga
D.Va from *Overwatch*, 2017. "I gave up this costume for 3 years due to the racist hate I received. Now I bring her back to motivate and inspire others." Also check out her Wakandan princess warrior: "When @cutiepiesensei asked me to be Shuri, I said yes before I even knew where to begin! It was ambitious." The bodysuit alone is sewn from 40 pieces.

VAGUELY COSPLAY OF UNDISCLOSED LOCATION @vaguelycosplay

Gamora from *Guardians of the Galaxy*, 2018. Also check out her newest cosplay, Black Canary from *Birds of Prey*, 2020. "I haven't done anything new lately because of everything that's going on in the world!"

Red Panda Photography

RYAN MOTTLEY OF HATTIESBURG, MISSISSIPPI @ligerzero_gaming

Sora from *Kingdom Hearts 3* takes a break from wielding the keyblade, 2019. Cosplayer, Twitch streamer, and video game writer.

R/C GARBAGE CAN ZOMBIE

Written by Mel Vallero, M.D.

Build the scariest prop in town and have more fun than your kids, with this drivable update of a classic

MEL VALLERO, M.D. is an emergency physician in Northern California and has been making, fixing, and MacGyvering since childhood. His background as an ER doctor and lifelong maker gives him the ability to fix almost anything. He's also the brother of Randell Vallero, M.D. ("DIY Fog Projection Screen," page 112).

TIME REQUIRED: A Weekend

DIFFICULTY: Moderate

COST: $300–$500

MATERIALS

» **Plastic garbage can with wheels and flip top lid** I used the Toter 32-gallon can.
» **Caster wheel, 2" diameter with single stem**
» **Galvanized metal strap, ⅛"×1½"×10"**
» **Motors, 12V DC, left-hand (1) and right-hand (1)** I used AM 218-series gearmotors, 116rpm with 64mm shafts, $107 each from robotmarketplace.com/products/0-218-1003.html. This project will be much less expensive if you already have some appropriate motors lying around or can get them "gently used."
» **Wheels with rubber tread, 8" (2)**
» **Radio control (R/C) transmitter and receiver, minimum 3 channel**
» **Dual motor driver, 12V DC, with R/C input** I'm using the Sabertooth Dual 12A 6V-24V R/C Regenerative Motor Driver.
» **Sealed lead acid battery, 12V 5Ah** like those found in household alarm systems or small ride-on electric toy cars
» **Battery quick-connectors**
» **Inline switch** for battery connector
» **Compressed air tank, 5 gallons**
» **Air regulator with gauge, ¼" NPT fitting** for the air tank
» **R/C relay switch** PicoSwitch Radio Controlled Relay
» **3-way solenoid air valve, 12V** with ⅛" NPT ports
» **Brass reducer fitting, ¼" NPT to ⅛" NPT** for solenoid valve
» **Compressed air cylinders, ¾" bore, 3" stroke (2)** with ⅛" NPT ports
» **Adjustable pressure valves (2)** ⅛" NPT, for each air cylinder
» **Various quick-connect couplers** ¼" NPT, for compressor and solenoid valve attachments
» **High pressure tubing, ¼", about 6' length** You can also use ¼" tubing for drip irrigation systems.
» **T-connector, ¼"** for pressure tubing
» **Various push-in tubing connectors, ⅛" NPT to ¼"**
» **Zombie mounting arm, 8"–10" long** Make it from a piece of wood, aluminum, or light tubular steel.
» **Strap hinge** for zombie mounting arm
» **Zombie prop** A skeleton head with small upper chest and arms, a plastic pumpkin head, or any scary small prop from your collection.
» **Various bolts, nuts, cotter pins, nails** for attaching motors, air cylinders, and zombie bracket
» **Angle, aluminum or thin steel, 8"×1"** for lid bracket
» **Mounting brackets (2)** for air cylinders
» **Shelf brackets, 2" (2)** for use with air cylinders; or use similar lengths of malleable steel
» **Wire, 10 gauge** for an improvised lid bracket hinge; see step 3
» **Motion activated spooky sound box (optional)** These can be found in specialty Halloween stores.
» **Low voltage wire** for electrical connections
» **Assorted double stick tape, tape, and zip ties**
» **Teflon tape** for NPT connections

SOURCES

» **amequipment.com, ebay.com** for motors
» **harborfreight.com** for compressed air tank, air regulator, ¼" NPT connectors, 8" wheels, caster wheels, tools
» **frightprops.com** for air cylinders, solenoid valve, ⅛" NPT fittings, tubing,
» **robotshop.com** for motor driver, R/C relay switch
» **homedepot.com** or **acehardware.com** for garbage can, steel and aluminum brackets, assorted nuts/bolts, fasteners, and tools
» **amazon.com** for R/C system, 12V batteries and everything else

TOOLS

» **Air compressor** for filling the compressed air tank
» **12V battery charger**
» **Screwdrivers**
» **Pliers**
» **Wire cutters**
» **Knife or utility blade**
» **Adjustable wrench or socket wrench set**
» **Hammer**
» **Metal file**
» **Keyhole saw or reciprocating saw**
» **Hacksaw and/or angle grinder with cut-off wheel**
» **Electric drill** with bits appropriate for steel
» **Center punch**
» **Cutting oil** for drilling into metal; or WD-40 in a pinch
» **Soldering iron and solder**
» **Tape measure, ruler, marker, pencil** the usual stuff to make things relatively straight and neat
» **Bench vise**

Halloween is a big deal at our house and every year it gets bigger! For better than a decade, we've been the scariest, noisiest, and spookiest house in town. Over the years, our haunt has evolved from a few pneumatic-powered zombie props on the front lawn to ghostly AV displays and multiple rolling, radio-controlled animated props trolling for victims up and down the street.

One of our all-time favorites is a mechanized device we affectionately call Oscar — a radio controlled, roving flip-top garbage can with a zombie prop that pops out with startling speed on command. It's powered by a small 12V sealed lead acid battery providing juice for the wheel motors, with a compressed air tank providing for

Mel Vallero, Adobe Stock - Vjom

Garbage can

Lid piston

Transmitter

Receiver, motor driver

Battery

Zombie piston

Compressed air tank

12V DC gearmotors

the rapid deployment of Oscar the zombie when unsuspecting trick-or-treaters are nearby.

We often park Oscar next to the sidewalk alongside some real garbage cans and let the fun begin. Alternatively, we slowly roll up behind unsuspecting kids and their parents and let Oscar pop as they turn around. Either way, there's plenty of dropped candy at the end of the night. We were fortunate to share Oscar and our other props with our local maker community at the 2018 Maker Faire Bay Area and it was nonstop fun all three days.

Oscar's "screaming success" comes from its ability to easily move from place to place, its initial benign appearance, and its very quick compressed air mechanism. It's a higher-tech update of the classic "Trash Can Trauma" prop, (makezine.com/projects/trash-can-trauma), and a worthwhile build if you're looking for the ultimate Halloween prop this year.

SCARY DRIVER

Oscar is made up of an electrical component that uses a 12V battery to power two electric motors for mobility of the garbage can, and a mechanical component that uses compressed air to power two air cylinders. One cylinder opens the garbage can's lid and the other deploys the zombie or scare prop of your choice.

Both electrical and mechanical components are remote controlled via a hobby R/C transmitter and receiver. The motors are driven via a dual motor driver board which plugs into the R/C receiver. The compressed air is released in short bursts via a 12V solenoid air

valve to activate the zombie and lid. This valve is controlled by a relay switch, which also plugs into the R/C receiver.

Although I refer to specific materials and methods, please use this tutorial only as a guideline. You may already have some of the necessary materials lying around in the garage, or you may have specialized skills like 3D printing, machining, and welding; adjust your build accordingly. The most expensive components are the motors which can set you back about $120 each if bought new. However, less expensive windshield wiper or car seat motors would work as long as they spin about 100rpm and their shafts are at least 2½" long to allow for wheel mounting. Of course, motor and wheel mounting will vary.

BUILD YOUR R/C GARBAGE CAN ZOMBIE
1. MOUNT THE MOTORS AND WHEELS

Remove the plastic wheels and steel axle from the Toter garbage can, and store them for other projects. The plastic wheels don't provide enough traction for this project and are more difficult to mount on the motor shafts. Besides, what maker couldn't find some other use for an extra couple of wheels and a solid steel rod?

Using a keyhole saw or reciprocating saw, cut a window with a flap on the bottom rear of the garbage can (Figure A) for installing the motors and accessing the electronics.

The specified drive motors have 3 equally spaced mounting holes that you'll use to bolt them to the thick plastic surrounding the garbage can's axle holes. To drill matching

Mel Vallero, Adobe Stock - Vjom, Caleb Kraft

holes on the can, first measure the center-to-center distance between two of the holes in the motor. On a piece of cardboard, draw an equilateral triangle with sides measuring this same distance and cut out a triangular hole. Center this triangular hole over the can's original axle holes and use a Sharpie to mark the three corners. These marks are where you'll drill holes; position them so that the body of the motors will end up lying somewhat against the bottom of the garbage can when mounted. Drill holes for each of the bolts at the three marks.

Push the motor shafts through the axle holes and secure each motor with three M6×40mm bolts with 1.0 thread (or whatever bolts fit your specific motors).

The drive motors have long shafts onto which each wheel can be mounted. Push a wheel onto each shaft and test-fit so that there's about a ¼" gap between the wheel and the garbage can wall; this allows for some sagging of the axles under load, otherwise the wheels will rub against the can. Using a ⅛" bit, drill through the plastic wheel collar (on the inside surface of the wheel). Continue until the bit begins to mark the surface of the steel motor shaft, then stop. Repeat this process on the other wheel.

Remove both wheels and motors to finish drilling through the hard steel shaft. Clamp the motor shaft in a sturdy vise with your drill mark facing directly upward. Use the center punch to make a dimple at your mark, then carefully drill a ⅛" hole entirely through the shaft. Use cutting oil or a little WD-40 when drilling through metal to keep the bit cool and sharp. Repeat this process for the other shaft.

Use the ⅛" bit to complete the drilling process through both wheel collars by drilling a hole opposite to the first hole drilled into the wheel collars. Reattach motors to the garbage can using the bolts; these may be tightened or loosened a bit to make the motor shafts horizontal on both sides. Slide the wheels onto each shaft, align the holes on the wheel collars and motor shafts, and use cotter pins or nails to secure the wheels (Figure ).

Cut and finish the steel strap for the caster wheel, using your hacksaw/angle grinder and file. Drill appropriately sized holes in the strap

(again using cutting oil) and garbage can. Bend the strap using the bench vise and hammer in a Z shape as shown, so that the garbage can travels relatively upright when the caster wheel is mounted. Install the front caster wheel and fasten the steel strap using nuts and bolts (Figure C).

2. CONNECT THE ELECTRICAL COMPONENTS
Connect the positive and negative leads of each motor to the appropriate motor driver terminals. Solder the inline switch in series to the positive

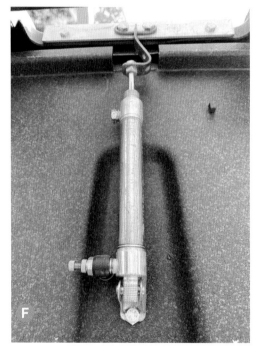

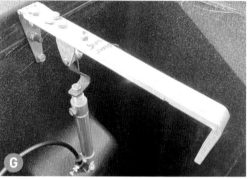

(+) lead of the 12V quick-connector and then connect it to the appropriate motor driver battery terminals. Plug the R/C leads of the motor driver into the R/C receiver unit, following the driver's instructions.

Mount the motor driver and R/C receiver against the inside wall or near the floor of the garbage can using double-sided foam tape (Figure D). Mount as low as possible, avoiding contact with the compressed air tank, battery, or other components. Mount the receiver antenna along one of the side walls of the can using double sided tape or duct tape. The battery should fit between the mounted motors on the floor of the garbage can; you can use stick-on

hook-and-loop fasteners to secure it. Solder a quick-connector to the battery leads.

Plug the R/C relay switch into the receiver. Using low-voltage connection wire, connect one terminal of the R/C switch to the positive lead of the battery (for ease, you may connect to the positive battery lead at the motor driver terminal). Connect the other terminal of the R/C switch to one of the 12V solenoid leads. To complete this circuit, connect the other solenoid lead back to the negative (–) terminal of the battery (or again connect at the motor driver terminal for ease). Now the solenoid will open the air valve when the R/C relay switch is activated (closed).

Double-check all connections. Make sure the transmitter has a fresh battery. Turn on the transmitter, connect the battery quick-connectors, and turn on the inline battery switch. On a two-stick R/C transmitter, the left stick should now control left, right, forward, and backward movement of the prop. You should also be able to hear the R/C relay switch and solenoid respond to their corresponding channel switch or stick movement on the R/C transmitter. When it's all working, turn everything off and go to the next step.

3. INSTALL AIR CYLINDERS AND ZOMBIE MECHANISM

Create the lid bracket using an 8"×1" aluminum angle or similar, as shown in Figure **E**. Drill evenly spaced holes for the bolts. I used ¼" bolts and nuts because I had a few lying around; #10 size or similar would work just as well. The lid bracket hinge was created with 10 gauge wire bent around two of the bolts, and a repurposed 2" shelf bracket which was drilled and bent; you can copy this or improvise your own hinge.

The air cylinder for the lid is attached to the outside rear wall of the trash can using a special "pivot mount" bracket, which can be purchased with the cylinder or created using a flat piece of steel and bent/drilled appropriately. Position the cylinder so the piston rod is fully retracted when the lid is fully closed. Use a utility knife to carefully cut a notch into the garbage can lip to allow clearance for the air cylinder push rod and bracket (Figure **F**).

Create the zombie mounting arm using a length of aluminum or wood about 8"–10" long depending on zombie size, and a small strap hinge (Figure **G**). I bent the end of the arm to prevent the zombie from flying off when fastened with zip ties, but this isn't necessary if you're using duct tape. The air cylinder pivot on the mounting arm was created using a cylinder bracket with a bolt serving as a hinge pin. The cylinder attaches to this hinge pin using another 2" shelf bracket bent and drilled appropriately. Attach the zombie cylinder inside the garbage can so that the piston rod is fully retracted when the mounting arm is fully lowered.

4. CONNECT THE COMPRESSED AIR COMPONENTS

Remove the short air hose that's attached to the 5gal tank. Using the Teflon thread seal tape to prevent leaks, install the air regulator with gauge to the compressed air tank and install the female quick-connect coupler to the air regulator (Figure **H**).

On the solenoid valve's inlet port, install the male end of the coupler using the ¼" to ⅛" NPT reducer. On the outlet port, install a push-in connector. If your valve has extra outlet ports, use screw-in plugs. A silencer may be used on the exhaust port of the valve if desired.

On each air cylinder, install an adjustable pressure valve to the port closest to the base. Push a length of the ¼" black pressure tubing into the solenoid outlet port connector. Push two additional lengths of tubing onto each of the adjustable pressure valves, then use the T-connector to connect these 3 tubes together. You'll have to drill a small hole through the rear of the garbage can wall in order to route the tubing to the cylinder on the back of the can.

After filling the compressed air tank using your air compressor, carefully lower it into the garbage can and connect the solenoid valve to the air regulator using the quick-connect coupler (Figure **I**).

Fully close the valves to both air cylinders and then open a half-turn initially (you'll adjust these in a bit). Carefully open the main valve on the air tank fully. If you hear air leaks, turn the air tank valve off and slightly tighten the compressed air fittings at the tank, regulator, and solenoid valve. If the quick-connect couplers seem to be the problem, Vaseline or light grease around these couplers will usually stop the leak. If no leaks, then you're good to go to the next step. Adjust the air regulator to about 40psi.

Stand back to test left/right/back/forward movement of the can. Refer to your specific R/C unit instructions to adjust neutral position, direction of movement, and sensitivity of the controls as needed. Be careful with fast starts, as the can may tip backward easily, especially when the 12V battery holds a fresh charge.

Flip the appropriate switch or stick on your transmitter to open the solenoid air valve. Adjust the two air cylinder valves as needed for desired opening speed. For smooth operation, the valves should be set so that the lid begins to open just a bit before the zombie is pushed upward. Once adjusted, use the lock collar on the valves to lock them in position, or simply tape them in place.

FINISHING TOUCHES

- For an added audio effect, use zip ties or double sided foam tape to secure an inexpensive small motion-activated spooky sound box inside the upper inner wall of the garbage can. The sudden opening of the lid should effectively trigger this sound box.

- If you have access to a Cricut machine or a laser cutter, make a decal for the side of your garbage can to match your real garbage cans. In a pinch, a paper logo affixed with double sided tape will be good enough in the relative dark of a Halloween night.

- For added realism, consider hanging a banana peel, candy wrapper, or other bits of trash over the edge of the garbage can.

ON HALLOWEEN NIGHT

Two batteries should last one busy Halloween night. Fully charge both before the festivities

ZOMBIE OUTBREAK!
SETUP AND TESTING

Use copious amounts of duct tape to fasten your zombie (or scary object of preference) to the zombie mounting arm. Make sure the transmitter joysticks and switches are off or in neutral. Turn the transmitter power on and then turn the 12V battery switch on (it's good practice to turn on the transmitter before the receiver in most R/C applications).

begin, and you can make one quick battery change in the middle of the night using the cutout and flap at the bottom of the trash can. You'll also have to refill the air tank once or twice, so have your compressor and hose ready.

Designate someone else to hand out candy at the front door while you walk around discreetly outside with the R/C transmitter. This way you can roll your prop up and down the sidewalk or street to meet your unsuspecting victims where they won't expect a scare.

CAUTION: TIPS FOR SAFE SCARING

- There's a potential pinch point between the rear garbage can handle and the pneumatic cylinder piston rod when it deploys, so never activate the zombie when anyone is holding this handle. Also, the zombie and lid can deploy quickly depending on air pressure and valve settings so avoid popping the prop when faces or hands are nearby.

- Avoid scaring very young kids (kids shorter than the garbage can should usually be off limits) or the very old. Your prop should only be used for good. My favorite targets are teenagers trying to look brave with their friends, or big "tough looking" guys holding a beverage or two.

- Your prop will look like a real garbage can, so be careful of those wanting to use it as such. However, these people also make some of the best victims ... Just pop the zombie before they toss the candy wrapper in the can and all is good.

- Avoid handing your R/C controller over to others during your haunt. You will likely not get it back!

UNTIL NEXT TIME

Before storing the prop, make sure all batteries are disconnected and removed, and all pressure is released from the compressed air tank. Store the garbage can so that it doesn't rest on the wheels, which could cause the axles to angle upward during long storage.

Have a safe, happy, and noisy Halloween! ⊘

FAUX FLAMING WINDOW
Written by Jaimie and Jay of Wicked Makers

Using simple box fans and orange lights to make hanging sheets flap like a raging fire, we turned our home into a Disney-esque Halloween attraction — for the neighbors and fire department alike. Here's how:
1. Hang a smooth sheet from a curtain rod and, from below, point a box fan upward to give the sheet a billowing effect.
2. Put an orange and yellow LED light bulb on either side of the fan to shine light upward onto the sheets.
3. Adjust sheet tension to get maximum flapping; when viewed from outside it's almost indistinguishable from real fire!

This simple special effect was inspired by the Pirates of the Caribbean ride at Disneyland. See the full DIY video and details at youtu.be/DustMUjfk58.

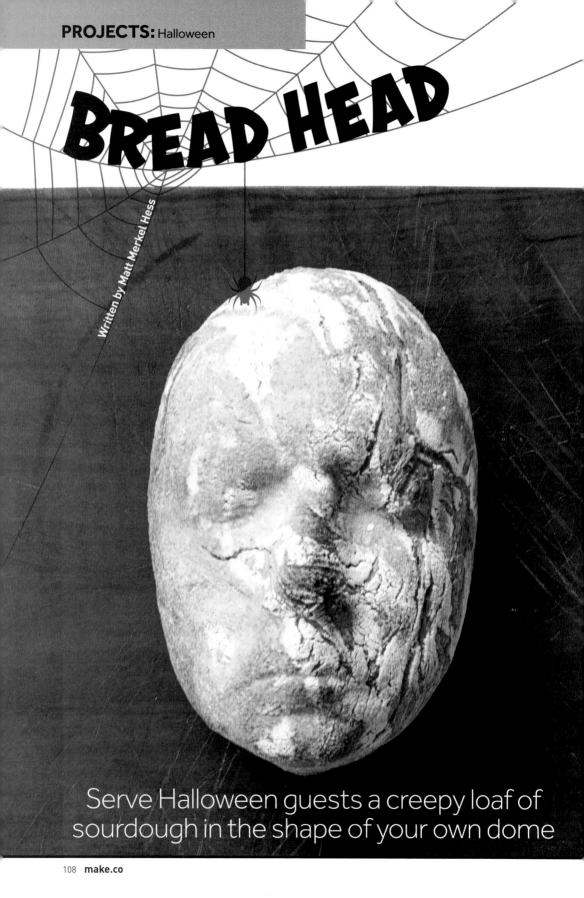

BREAD HEAD

Written by Matt Merkel Hess

Serve Halloween guests a creepy loaf of sourdough in the shape of your own dome

TIME REQUIRED:
A Few Weeks

DIFFICULTY:
Easy–Moderate

COST:
$60–$70 plus $10–$30 for kiln firing

MATERIALS

» **Smooth-On Alja-Safe alginate,** about 1lb
» **U.S. Gypsum No. 1 Pottery Plaster,** 10lbs
» **Gypsum bandages** aka plaster gauze, to support alginate cast
» **Low-fire red terracotta or earthenware clay,** 5–10 lbs
» **Bread-making supplies: Flour, yeast, salt, water** Your favorite recipe should work.

TOOLS

» **Buckets** for mixing alginate and plaster. Make sure they're clean.
» **Drill and mixer blade or mixing paddle** for mixing alginate and plaster
» **Poncho or plastic sheet** to protect clothes during face cast
» **Hair gel or petroleum jelly** to protect eyebrows
» **Clay sculpting tools** such as plastic or metal ribs
» **Kiln access** to fire your terracotta baking dish

MATT MERKEL HESS is an artist, educator, father. Midwesterner in New York City via Los Angeles. Find him at merkelhess.com.

I made these self-portrait baking dishes for a 2014 fundraiser at the haunted Greystone Mansion in Beverly Hills, to benefit the nonprofit **LAXART.** I was trying to make something that was a bit spooky and interactive. The final project was presented in the kitchen, where the bread was sliced and toasted, and then guests could add butter, jam, and honey. It was a really fun project, and I enjoyed each step of the way. Here's how to do it.

1. PLASTER-BACKED ALGINATE FACE CAST

First you need to make an alginate cast of your face. The best product for this is Alja-Safe alginate from Smooth-On, which is nontoxic and designed to be used with body and face casts.

You'll need a helper for this. Cover your eyebrows and any facial hair with hair gel or petroleum jelly, and put on a plastic poncho or something to cover your clothes. Next, mix the alginate in a clean bucket. Your helper will then apply the lavender-colored alginate to your face and let it set (Figure Ⓐ). Be sure to leave your nostrils clear (Figure Ⓑ), and you can also keep a straw in your mouth to breathe. When I made the cast, my face was "relaxed" and the cast ended up as a sort of death-mask frown (Figure Ⓒ). So think about the expression you want in your mold, and also an expression that you can hold throughout the 30 minutes or so it will take to make this mold.

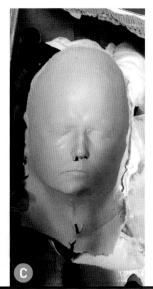

Ⓐ   Ⓑ   Ⓒ

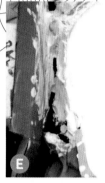

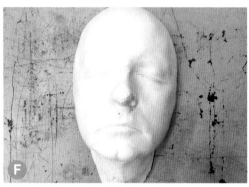

Matt Merkel Hess, Adobe Stock - Vjom, Caleb Kraft

After about 10 minutes, reinforce the cast with plaster-gauze bandages. Add salt to the water for dipping the bandages to speed drying time. In Figure D, you can see I'm also holding a hair-dryer to speed the drying process.

After the plaster gauze stiffens, carefully demold the cast from your face. Alginate starts to break down within 12 to 18 hours; best practice is to make your next mold within 4 hours.

2. PLASTER POSITIVE

The plaster-backed alginate cast should be quite stiff, but it can be backed with more clay so that it doesn't bend or deform. Putting this all in a cardboard box will give you extra support and help contain any plaster drips.

After preparing your mold, mix plaster according to the manufacturer's direction and pour it into the mold to cast a positive (Figure E). You can use any quality plaster from U.S. Gypsum; I mix No. 1 Pottery Plaster at 10 parts plaster to 7 parts water by weight. I wouldn't use plaster of Paris, as you want a really fine cast that will capture all the detail.

After the plaster hardens (about an hour),

remove it from the alginate mold. If necessary, plaster can be scraped, cut, and sanded, so this is your chance to get the mold how you want it (Figure F and G). You might be able to use the alginate cast again, but the heat of the plaster curing can alter the alginate.

The plaster cast needs to dry. This takes 1 to 2 weeks by air drying, or you can speed it up by putting it in a low oven or hot box. Make sure to keep cured plaster under 150°F or you risk ruining your mold.

3. TERRA COTTA BAKING DISH

After the plaster positive is dry, the next step is making the baking dish. I used a traditional low-fire red terracotta clay, but you could use just about any water-based clay.

> **NOTE:** Air-dry clays and polymer clays like Sculpey can't be used in the oven to bake bread, so don't use them.

You will also need access to a kiln; most community studios will allow you to fire by paying by volume, typically 3¢–8¢ per cubic inch. If your dish is around 10"×6"×6", it'll cost about $11–$29.

H

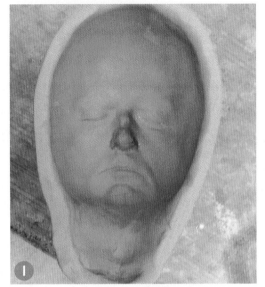

I

To make your baking dish, roll out about 5lbs of clay into a ¾"-thick slab, then press the slab over the plaster positive. Try to use one single slab of clay to cover the entire mold, so you don't have a crease inside that needs to be cleaned up. Smooth the clay and press it onto the mold, so that you get every detail of the plaster. Do your best to keep the thickness even throughout. Then, add feet (Figure H) or a footring so the baking dish will sit evenly in the oven. After the clay has dried a bit and can support itself, carefully remove the clay from the plaster (Figure I).

Next, dry the baking dish completely and fire in a kiln to cone 04 temperature, or about 1,950°F. Clay shrinks, so your final baking dish, or baker, will be about 6 percent smaller than life size.

4. BAKING BREAD

Place the baker in the oven and preheat to 450°–500°F. Once the dish is hot, you can drop about 750–1,000 grams of fully proofed dough inside and bake (Figure J). The dough needs to be flexible enough that it will take on the shape of the nose and lips, so a higher-hydration recipe is recommended.

It's key to preheat the baker so that the bread receives even heat from all sides. Bake until it's golden brown on top, about 40 minutes, and remove (Figure K).

You may need to use a butter knife to pry the loaf away from the side, but it will eventually pop out and you've got self-portrait bread. Let cool and enjoy!

This project was really fun to do, and the result is a baking dish that can be used over and over again. My kids love these loaves, and every so often, I'll drop an image of the "head bread" on social media and enjoy the comments. If you're not into self-portraits, this technique could be used to make any sort of baking dish form. Just make sure there are no undercuts or protrusions that will interfere with the bread slipping out of the mold. I look forward to the delicious results from your efforts. ❂

J

K

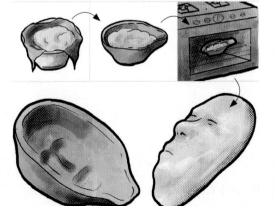

DIY FOG PROJECTION SCREEN

Low-cost and lightweight, this indoor-outdoor screen looks awesome and adapts to the weather

Written by Randell Vallero, M.D.

RANDELL VALLERO, M.D. is a physician working in gastroenterology who's interested in all sorts of fabrication, including 3D printing. Prompted by PPE shortages, he and his son designed snorkel mask respirators, featured in *Make:* Volume 73.

Projecting video on a fog screen creates an awesome ghostly effect. It's not a novel idea in 2020; there are commercial fog screens for $10,000 (I did find one on Alibaba for $4,000) and some excellent DIY fog screen projects posted online. But none of these met the criteria I wanted; they were either too expensive, too heavy (400lbs!), or too impractical for a simple haunted house effect. I needed a fog screen that was:

- Cheap to build, from everyday materials that are easy to source
- Lightweight enough to hang over Halloween revelers' heads without a scaffold
- Adaptable to indoor/outdoor conditions such as warm windy days, cool calm days, or rain
- Easily deconstructed for storage (in a home that's already maxed out).

Such a fog screen didn't exist, so I decided to make my own. In this build I'll demonstrate how to make the cheap, light, highly adaptable fog screen I use in our yearly Halloween display, for less than $150 (not including your projector or fog machine). You can get all the electronics on Amazon and the rest from a hardware or home improvement store.

1. BUILD THE FOG SCREEN

Drill ⅓" holes spaced 1" apart in a straight line along the PVC pipe (Figure Ⓐ). This is where the fog will come out. You can use the entire length of the tube but I found that a 5'-wide fog screen is adequate. Cap one end with the PVC end cap.

For the screen fans, line up the 120mm computer fans evenly on either side of the holes you just drilled and secure them with zip ties (Figure Ⓑ). Orient them so the flow of air is in the same direction as the fog exiting the tube. Connect them in parallel by plugging the 4-pin Molex connectors together front-to-back to make one long daisy chain (Figure Ⓒ). Take the last fan's 3-pin CPU connector and use jumper wires to connect its motor terminals to one of your DC potentiometers. This will control the speed of all the screen fans on either side of the holes.

For the fog suction fan, sand the corners off the small 50mm fan to fit it inside the PVC pipe, with the flow of air directed toward the fog exit holes (Figure Ⓓ on the following page). Connect this fan to your second potentiometer.

Randell Vallero, Adobe Stock – Vjom

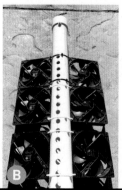

TIME REQUIRED:
A Weekend

DIFFICULTY:
Easy–Intermediate

COST:
$100–$150

MATERIALS

- » **Computer fans, 120mm (10–14)** I used Rosewill ROCF-13001, Amazon #B00KB8CB9O.
- » **Computer fan, 50mm** Amazon #B00006B8CM
- » **Potentiometers, 12V DC (2)** such as Amazon #B010JLKUMW
- » **Transformer, 12V** Many with various amperage ratings can be found on Amazon (see step 1).
- » **PVC pipe, 2" diameter, 6' length**
- » **PVC end cap, 2"**
- » **Eye bolts, 2¼"×3" (2)**
- » **Corrugated plastic panel, 4mm, 1 sheet** such as Coroplast. You need about 4"×150" total length.
- » **Zip ties, 8", 1 pack**
- » **Flexible duct dryer hose, 3"×20'**
- » **Bucket, 5gal, with lid**
- » **Dry ice, 20lbs**
- » **Fog machine, variable/continuous output** such as the Chauvet DJ H1200, Amazon #B014PHJEU0
- » **Video projector**
- » **3D printed connectors (optional)** Free files thingiverse.com/thing:4433607. Or use duct tape!

TOOLS

- » **Duct tape**
- » **Hacksaw**
- » **Drill with ⅓" drill bit**
- » **Hole saw, 3"**
- » **3D printer (optional)**

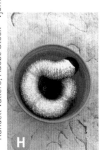

Connect both potentiometers to a 12V DC transformer that's rated for a total amperage greater than the sum of the amperage of all your fans combined. In our fog screen we have 14 fans at 160mA and one at 100mA, totaling 2340mA, powered by a 12V 2500mA transformer. We were lucky and sourced the transformer from an old laptop but various 12V transformers with the needed amperage can be found on Amazon.

Finally, make the fog skirt. Previous DIY and commercial fog screens have focused on laminar flow of the air from the fans to keep the fog in place. That's awesome, but it adds too much weight and complexity for our needs. I found that a 4"-tall rectangular skirt made from corrugated plastic, running the length of the fog tube under the exit holes (Figures **E** and **F**), helps focus the fog on a thin line after exiting. The skirt can be secured with duct tape or with our 3D-printed supports.

Eyebolts can now be drilled into each end of the pipe to suspend the screen with nylon cord (Figure **G**) without the need for scaffolding. The entire fog screen tube with fans and electronics weighs a mere 8½ pounds!

2. FOG COOLING BUCKET (OPTIONAL)

Ideally the fog should be cooled so it behaves "heavier" as it leaves the PVC tube and disperses less, which leads to a better projected image. But I have successfully used the fog screen without cooling the fog, by varying the speed of the fans. For a quicker build, this component may be skipped if necessary.

We run our fog through a simple 5gal cooling bucket. Cut two 3" holes in the bucket wall, at the top of one side and the bottom of the other. I made these holes 180° from each other. Then coil 10' of 3" aluminum dryer hose from the bottom of the bucket to the top, exiting the holes and leaving the center empty (Figure **H**). Fill the bucket with dry ice for rapid cooling and put the lid on. Connect the top opening of the cooling bucket to the suspended fog screen with the remaining 10' of dryer hose. All connections can be made with duct tape or our 3D printed parts.

Connect the fog machine to the lower opening of the cooling bucket and your fog screen setup is ready to go (Figures **I** and **J**).

NOTE: For a bigger chill, try Adam Tourkow's "Ultimate Fog Chiller" (makezine.com/projects/ultimate-fog-chiller), featured in the *Make:* Halloween special editions of 2007 and 2016 (makershed.com/pages/halloween-pdfs).

3. FOG MACHINE

Don't skimp. After much experimentation, I've found it's crucial to get the right amount of fog for the situation. On a windless day with hyper-cold fog cooled by dry ice, you may need very little fog. On a windy day when you don't have dry ice and the fog is warmer, you may need to output more fog. The fog must also run continuously. Most cheap fog machines run 100% for 40 seconds and then stop and have a refractory period before they fire up again; this is not acceptable for our fog screen. The best fog machine for the screen would be one with variable output and also on-demand output (Figure **K**). I found the Chauvet DJ Hurricane 1200 to be the best and least expensive fog machine that fit these criteria.

4. PROJECTING VIDEOS AND IMAGES

The final step is projecting onto the screen. Ghostly images and video loops can be found online, both open source and for purchase. A single ghost or phantom without any background usually projects very well (Figure **L**). I personally like the quality of the video loops at AtmosFX.com but you can project anything that you would project on a real projector screen.

Good projections can be made even with average projectors. Wait until dark, and try blocking unwanted ambient light near the screen, and varying the fog output and the fan speeds. For best results the projector must be behind the screen relative to the viewer, and after much trial and error, I've found that placing the projector at a higher elevation than the screen tends to result in less glare for the viewer.

The heavier the fog, the sharper the resolution of the projection, with crisp images and detail; but too much fog can fill an area quickly, ruining the effect. Lighter fog done just right is almost invisible, and the projection seems to materialize out of thin air like a hologram. Detail is harder to see with the lighter fog, and if it's too light the projection may not be seen at all. It takes practice,

THE DIY FOG SCREEN

Suction fan — Screen fans — PVC fog emitter — Fog skirt — Video projector — Fog machine — Dryer hose — Fog cooling bucket

varying the fog output, fog suction speed, and screen fan speed to achieve the desired illusion for the weather conditions that day.

FRIGHT NIGHT TESTED!

Our fog screen has successfully entertained and scared many, for several Halloweens in varying weather including windy, warm, cool, and rainy nights. It even made an appearance indoors in the Dark Room at Maker Faire Bay Area 2018. With its lightweight construction and ability to modulate the various settings to account for conditions, it fills a new niche among fog screen builds.

I'd love to see your adaptation. Please share pics and questions @qebehsenuef. ●

D-I-PLY

Reuse hollow-core doors to make plywood and more, with less work than you think

Reusing isn't new, but items deliberately made from reclaimed materials have become statement pieces. The "upcycling" movement and rising interest in eco-conscious wares has given value to the valueless. Decrepit and dangerous barns once left to decompose are now painstakingly picked for profit. Even shipping pallets have earned a place as sought-after sources of wood. And then there's my latest obsession: hollow-core doors (Figure A).

You know what I'm talking about: Those flat, thin doors in casa de Brady Bunch. Inexpensive, (then) stylish, lightweight, and surprisingly durable, they did the job. Alas, times change and nowadays when someone remodels a home these doors are often the first to get tossed. I would guess tens of thousands go to landfills annually. But wait, they have tons of value! I've set out to save as many doors as possible and put them back in service. Here is the abridged version of everything you need to know about reclaiming hollow-core doors.

WHAT ARE THEY?

Hollow-core doors typically consist of two ⅛" plywood-skin sheets (one per side), a 1"-thick solid wood frame (usually pine), and a honeycomb of corrugated cardboard filler to add structure to the centers. They are usually about 1¼" thick in total. The wood frame is about 2" deep on the top and bottom so there's real wood to cut into when adjusting a door to a jamb, and about 1½" on the sides, to hold hinges. Additional 2" "knob blocks," typically 12"–18" long, are added to the inside edge of the frame on both sides, so a door knob can be mounted on either side. I have also seen doors with thin strips of wood inside them

and recently found a batch with slices of thick cardboard tubing used for interior structure.

FINDING DOORS

Besides keeping your eye open at the curb, almost any unremodeled house built between the 1950s and 1990s is full of them. If a house from that era is being renovated in your area, ask! Most contractors will gladly give you the doors to save the dumping fee. Even better, tell your local contractors you want them and they'll call you from each new job.

I frequently put out messages in local Facebook groups and other social media sites. I also asked friends and family to spread the word and have never had a problem finding someone with a basement full of doors or a home improvement project they've been putting off. One time my friend Peter Kelly of 542 Woodworks had just finished a big renovation job and I scored more than 40 of these doors destined for the landfill in one trip (Figure B)!

A

B

TIM SWAY is a Connecticut-based artist and maker who specializes in reclaimed, upcycled, and eco-friendly woodworking. His mission statement is to "Make Worthless Things Priceless" and he's currently focused on making affordable and eco-ethical guitars. You can learn more about him at youtube.com/TimSway, NewPerspectivesMusic.com or timsway.net.

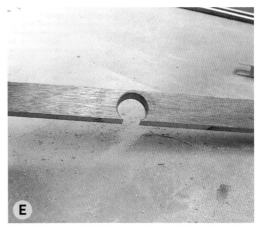

DISMANTLE A DOOR

If a hollow-core door is 30"×80", expect to get two sheets of plywood around 23"×75", plus some usable pine strips.

The doors are glued together with something like hot glue, so the cardboard parts come off fairly easily but if you're greedy and try to pull the thin plywood from the wood frame you'll almost always wreck it. Cut your losses on the wood frame around the edge and cut it off (but don't worry, we're still going to use that wood, too).

Start by cutting the long sides off (about 1½" from the edge is usually enough) (Figure **C**) and put those aside. I typically use a table saw for this, and cut through the whole door in one pass, but a circular saw works too. Now you can see the top, bottom and knob blocks inside. Cross-cut the top and bottom off (use the circular saw and some masking tape to keep these cuts clean) (Figure **D**). Then cut off the knob blocks (Figure **E**). The easiest way is to slice the whole door another 2". You'll end up with strips of the plywood on either side of the block you can snap off and save for smaller things (Figure **F**).

Sometimes you can get the plywood off the knob blocks and get a wider sheet about 27"×75" but the risk of wrecking it increases. I often will cut off only one knob block and take my chances with the second.

Once all the solid wood is removed, you can peel the two layers apart (Figure **G**). Wear gloves, as the glued corrugated cardboard edges are sharp. Then I usually use a wooden mallet to bang off the stuck cardboard (Figure **H**) rather than a scraper, which I find too time consuming.

Another option is to stand the door up tall and use a piece of wood or metal the same width as the interior of the door to push the glued cardboard off both sides at the same time. It's a little more of a cardio workout but you can do both jobs in one motion (Figure **I**).

At this point, if you've dismantled one 30"×80" door, you should have two strips of pine approximately 1"×1½"×80", two pieces of pine 1"×2"×27" and two pieces 1"×2"×18" (one with a hole in it), as well as two 23"×75" sheets of ⅛" plywood, four plywood strips about 2"×30", and very little waste besides the cardboard (which can be recycled). Well done!

USE AS-IS

I love having this ⅛" plywood around the shop (Figure J). It is the perfect material for cutting templates, making shims, making drawer bottoms, and more. Almost any idea I have gets prototyped in door skins first (Figure K). It's also great, free material for laser cutters. I made an entire acoustic guitar from doors on my laser (Figure L).

DIY PLY

To make thicker plywood from door skins, simply glue layers together to your desired thickness. The only downside is anything thicker than ¼" will require you to sand both sides of the skins to remove any cardboard glue residue from the inside and finish from the outside. This is the worst part of the job, but if you use something coarse like 60 grit sandpaper to start, you can make pretty quick work of it.

Use a generous amount of wood glue and make sure every square inch is covered (Figure M). I used spring clamps on the edges (a lot of them) and as much weight as I could find to put in the middle to clamp it up (Figure N). I've been pleased with the results every time with no voids (Figure O).

LAMINATIONS

But why not keep gluing? Especially since some of your conquest is cut into smaller pieces, why not create full laminated blocks of wood from the plywood (Figure P)? Alternate directions, create

patterns, get creative! I've made a few guitars and other things out of scores of skinny layers of plywood glued up and planed into interesting engineered blanks (Figure Q). When you cut and sand these at angles, all sorts of interesting wavy patterns emerge.

OTHER USES

While the plywood is certainly the point of the exercise, there's still a lot of good wood in the edges that can be used. The trim and knob blocks are usually Douglas fir or some other pine, but I've found most of this wood to be pretty high quality and, as a bonus from being hung in a door for decades, incredibly straight and dry — better than any green pine you'll get at the big box store.

Sometimes I run it through the table saw to remove the plywood still stuck to it and get nice, raw pieces of wood, but sometimes I leave the plywood on there to create new, unique looks (Figure R).

> **WARNING:** There may be little brad nails and/or metal staples in the frames that you will cut through. Your table saw blade will be OK but you may choose to use an older or less expensive blade. And, of course, wear proper PPE.

Hopefully you will never look at an outdated closet door the same way again — and maybe you'll find some inspiration to add them to your stockpile. Now that we've added these doors to the save list, what's next? ◢

USE THE HASHTAG

#HollowCoreDoorsAreTheNewPallet to find more ideas and to share your creations, and peruse youtube.com/TimSway channel for more information. I have more than a dozen builds featuring these doors as well as an in-depth video about my dismantling technique.

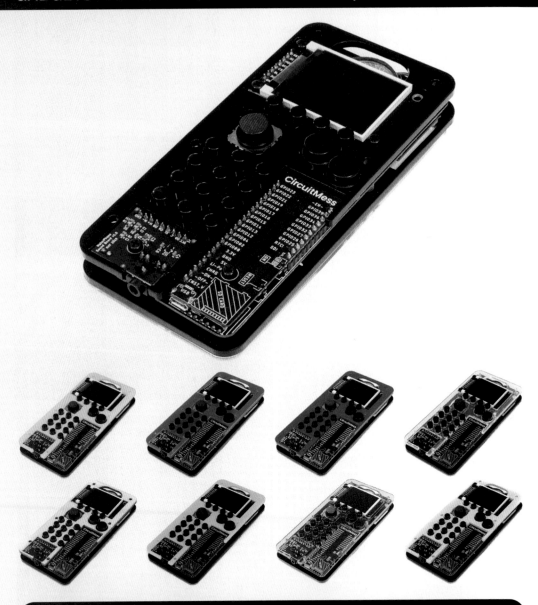

RINGO THE DIY MOBILE PHONE KIT

$170 circuitmess.com

This is a very interesting kit. It isn't particularly complex to assemble, just a fair amount of standard through-hole soldering, as well as a few screws. However, what you accomplish with the build is a fully functional cell phone that also acts as a programmable platform for apps and games.

You might be picturing a modern smartphone with a high resolution display, but this is closer to a GameBoy color than an iPhone. That's not a bad thing though. The tactile feeling of the phone in your hand and the solid click of the buttons along with the low resolution graphics all mesh quite well. —*Caleb Kraft*

HYPERTHERM POWERMAX30 AIR PLASMA CUTTER

$1,939 hypertherm.com

Plasma cutters are so cool, but the ones I've used all required hookups to large compressors, along with added moisture traps and flow valves. So when I got to test this unit with its built-in compressor, I worried that it would struggle to put out enough pressure to do long jobs. This concern grew when I saw the unit's small size — it's a bit smaller and lighter than your standard welding unit. Happily, it delivered impressive results with minimal setup: Just plug it in, attach the grounding clamp, and pull the trigger. The internal compressor kicks in instantly and, in my use, went as long as I needed it. The unit automatically switches between 120v and 240v (and includes tails for each plug type), and with up to 30amps of power it can cut metal ⅝" thick.

One realization though: After years of becoming accustomed to CNC precision, I really notice that my handheld plasma cutting is pretty rough. But this unit is compatible with CNC plasma tables, which we'll test out soon. This is a professional-grade machine with a matching price tag, but if you need portable plasma cutting, it's got you covered. —*Mike Senese*

ORANJE DESKTOP FOAM CUTTING MACHINE

$392 polyshaper.eu

Hot wire foam cutting has been a staple of industrial applications for many years. Polyshaper is bringing this ability to the desktop with their line of simple CNC foam cutters. The Oranje (the smallest in their lineup) offers a reasonable price to jump into cutting foam.

You really need to think in 2D when designing for this, as it is essentially only capable of cutting through a single plane, but it does that quite well. The hot wire slides effortlessly through foam materials of varying thickness; it sliced up 2"-thick foam without issue in my workshop.

Though you're stuck cutting in 2D, some creative design allows you to use this as a rapid prototyping tool for 3D shapes built from stacked or interlocking parts. A great addition to your rapid prototyping tool set.

—*Caleb Kraft*

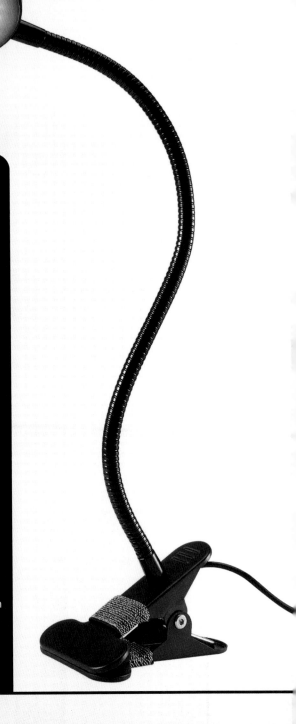

IVICT 24 LED USB BOOK CLAMP LIGHT

$23 amzn.to/2CzbOpR

Most of us are spending a great deal of time on video conferences, or getting more and more into hosting our own videos these days. My home office is pretty dark, which does me no favors on camera, so I sought out a reasonably priced ring light to help brighten up my face.

Although this gem is labeled as a book lamp, it works great for my camera needs and doesn't bother my eyes in the least — they actually made this lamp with kids in mind. You get a choice of light too: cool, warm, or a mixture of both, along with 10 different levels of brightness for each mode (30 lighting options)!

The sponge-lined clamp protects the surface you hook it to, and with an adjustable gooseneck stem you're able to set it up and readjust it to any angle on the fly. The view through the top makes it great for working on miniature projects on your desktop, too.

My only complaint is the clamp opens just 2" wide, so before you buy, make sure it will work on your ideal location. It just barely fit on the edge of my Apple monitor.

—Juliann Brown

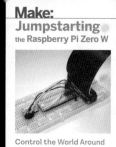

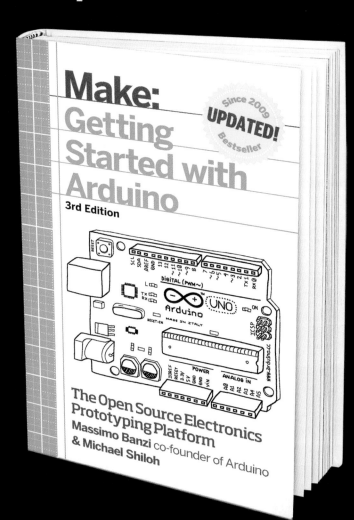

FARMCURIOUS FERMENTING SET

$30 farmcurious.com

If you're getting into fermenting, an airlock lid makes the process a lot cleaner and easier by keeping out the atmospheric elements that lead to mold. There are lots of options for this on the market; I like the set from Farmcurious because of its flip-close cap that lets you turn your mason jar into a pourable storage container when done.

The process is simple: Just fill up the container with your veggies, add some salted-water brine, put on the water-filled airlock, and store for a few days or weeks (depending on what's inside), and enjoy your own fermented goodies. Sauerkraut, kimchi, sriracha, and so much more, all made the best way: by you.

The Oakland, Calif.-based company also carries a variety of other useful and eye-catching kitchen and garden items, and hosts local and online classes. Check out their website for more info. —*Mike Senese*

I MAKE
by Tracy Blom

$12 amzn.to/3eyo3RG

This is a children's book, no doubt about that. Uniquely, however, it is a children's book that carries a maker-oriented message with a lead character with whom we're all familiar, Jimmy DiResta. Based loosely on some of Jimmy's experiences, this book is a delightful collection of pleasant illustration (done by Dahn Tran), well crafted story, and even some nice crafts to boot.

The story follows along as a young maker (DiResta) earns the respect of the neighborhood kids by showing them how he can fix and build stuff that they previously only considered buying. Great lesson here and adorable delivery. —*Caleb Kraft*

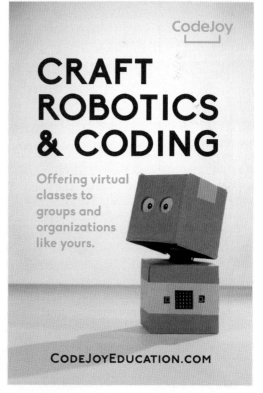

OVER THE TOP

Universal Lego Sorter

Built entirely from bricks, this machine can identify and sort any Lego part made, ever

DANIEL WEST
is an AI and computer vision software engineer based in Wollongong, Australia.

Ever since I saw Akiyuki's Lego-sorting machine way back in 2011, I'd had the idea of building one as well. About 3.5 years ago I was in university for computer science, and started playing around with computer vision. About a year later, I started reading some articles on AI, and realized it was perfectly suited for the task.

My machine first separates the bricks into a stream, using belts and a vibrating table powered by a Lego motor spinning an offset weight. This part of the machine went through more iterations than any other component, and even then I'm not totally happy with the solution — sometimes parts will fall through more than one at a time.

The separated parts then move through a "lightbox" that records video for AI classification. Mechanically this is the simplest part of the machine — all the hard work is left to the AI.

The parts are then distributed into 18 output buckets, using a series of gates controlled by servo motors. I've never seen anything like this done before, and it means that the machine can

be used for self-contained Lego storage.

The most difficult and complex part of the project is the software. A Raspberry Pi has to interpret individual frames of the scanner's video to extract cropped images of the parts. Those images are sent to the AI convolutional neural network, which is trained to detect the part number out of almost 3,000 possible options. The neural network is trained using 25 million "synthetic" images of digitally rendered Lego bricks. To improve accuracy, the network is also trained on about 200,000 "real" images, which had to be manually labeled with their correct part numbers.

The video I've released has inspired lots of others to start working on their own machines, and I'm very excited to see how people will improve and refine the technology! —*Daniel West*

 See Daniel's Lego sorter in action:
youtu.be/04JkdHEX3Yk